Code of Practice for Project Management for Construction and Development

Fifth Edition

CIOB
THE CHARTERED INSTITUTE OF BUILDING

WILEY Blackwell

This edition first published 2014
© 2014 by The Chartered Institute of Building

Registered Office
John Wiley & Sons, Ltd, The Atrium, Southern Gate, Chichester, West Sussex, PO19 8SQ,
United Kingdom.

Editorial Offices
9600 Garsington Road, Oxford, OX4 2DQ, United Kingdom.
The Atrium, Southern Gate, Chichester, West Sussex, PO19 8SQ, United Kingdom.

For details of our global editorial offices, for customer services and for information about how to
apply for permission to reuse the copyright material in this book please see our website at
www.wiley.com/wiley-blackwell.

Library of Congress Cataloging-in-Publication Data

Code of practice for project management for construction and development. – Fifth edition.
    pages   cm
  Coordinated by CIOB.
  Includes bibliographical references and index.
  ISBN 978-1-118-37808-3 (paperback)
1. Building–Superintendence.   2. Project management.   I. Chartered Institute of Building (Great Britain)
  TH438.C626 2014
  690.068'4–dc23

                                        2014017295

A catalogue record for this book is available from the British Library.

Wiley also publishes its books in a variety of electronic formats. Some content that appears in print may not be available in
electronic books.

Cover photo courtesy of iStock Photo
Cover design by Steve Flemming at Workhaus

Set in 10/13pt Franklin Gothic by SPi Publisher Services, Pondicherry, India
Printed and bound in Singapore by Markono Print Media Pte Ltd

1   2014

Contents

CHAPTER

4

Pre-construction 217

CHAPTER

5

Construction 251

CHAPTER 6

Testing and commissioning   263

CHAPTER 7

CHAPTER 8

Foreword

The first edition of this *Code of Practice*, published in 1992, set out a job specification for a project manager and provided guidance on the project manager's role. Since then project management has become an integral part of the construction industry and been responsible for its increased reliability and quality of product.

The next few decades experienced some significant changes within the industry with much focus towards changing our culture and communication. The interaction between the key participants in this industry, which produces many spectacular projects with increasing levels of complexity and technological prowess, continue to evolve around the necessity to deliver projects within an agreed budget, to a level of acceptable quality and within an agreed time scale.

The fourth edition, published in 2010, captured a range of themes across the industry. In this fifth edition, prepared in collaboration with a number of key professional bodies, the entire document has been overhauled to make it more contemporary while maintaining the integrity and rationale of the role of a project manager and project management in context of the construction industry.

Following the spectacularly successful delivery of the Olympics (London 2012) and continuing with the UK Crossrail project, construction is at the forefront of successful project management. This fifth edition, although developed specifically for the UK construction industry, will continue to satisfy the ever increasing demand for an authoritative document on this subject in other parts of the world.

I strongly commend this valuable multi-institutional code of practice to all the industry's clients, to practising project managers and indeed to all students of the subject and their mentors.

Jack Pringle, PPRIBA Hon AIA FRSA DipArch BA (Hons)
Principal, Managing Director
Pringle Brandon Perkins+Will

Acknowledgements

The fifth edition of the *Code of Practice*, under the stewardship of David Woolven FCIOB, has strived to keep pace, and in places perhaps steer the directions ahead, in the construction industry which has been at the centre of economic regeneration and development across the globe.

In keeping with the fourth edition, the fifth edition has also been prepared by a broad representation of the industry, with contributions from built environment specialists and interdisciplinary cooperation between professionals within the built environment. I would like to take this opportunity to thank the many people who have helped with the fifth edition. A list of participants and the organisations represented is included in this book.

Specific note of thanks must go to Piotr Nowak, who has been ably and patiently assisted by Una Mair throughout the delivery process, for coordinating all the disparate elements of the review of the *Code of Practice* by maintaining the information flow and also for managing the digitalisation of all the figures and diagrams.

I would also like to thank Arnab Mukherjee, FCIOB, who led the editorial and drafting team, for the successful delivery of this document.

Chris Blythe
Chief Executive
Chartered Institute of Building

Working group for the revision of the *Code of Practice for Project Management* – Fifth Edition

Saleem Akram, BEng (Civil) MSc (CM) PE FIE MAPM FIoD EurBE FCIOB	Director, Construction Innovation and Development, CIOB
Colin Bearne	Gardiner & Theobald
Sarah Beck MRICS MAPM	Royal Institute of British Architects
Andrew Boyle	Tesco
Shaun Darley	Voice of Reason Ltd/MB PLC
John Eynon	Open Water Consulting
Dr Chung-Chin Kao	Innovation & Research Manager, CIOB
Una Mair	Scholarships & Faculties Officer, CIOB - Group's Secretary
Gavin Maxwell-Hart BSc CEng FICE FIHT MCIArb FCIOB	CIOB Trustee
	Institution of Civil Engineers
Alan Midgley	ARUP
Arnab Mukherjee BEng(Hons) MSc (CM) MBA MAPM FCIOB	Technical Editor
Paul Nash MSc FCIOB	Turner & Townsend
Piotr Nowak MSc Eng.	Development Manager, CIOB
Dr Milan Radosavljevic UDIG MIZS-CEng ICIOB	University of the West of Scotland
Eric Stokes MCIOB FHEA MRIN	Salford University
David Woolven MSc FCIOB	Chair Working Group
	University College London
Roger Waterhouse MSc FRICS FCIOB FAPM	College of Estate Management, Royal Institution of Chartered Surveyors, Association for Project Management

The following also contributed in development of the fifth edition of the *Code of Practice for Project Management*

Andrew Barr	Davis Langdon
Richard Biggs MSc FCIOB MAPM MCMI	Construction Industry Council
Richard Humphrey FCIOB FRSA FCMI FIoD MAPM PGCert FHEA EurBE	Northumbria University at Newcastle
Vaughan Burnand	Chair, Health & Safety Advisory Committee
Professor Farzad Khosrowshahi FCIOB	Head of School of the Built Environment & Engineering Faculty of Arts, Environment & Technology, Leeds Metropolitan University
Dean Hyndman	URS
Dr Sarah Peace BA (Hons) MSc	Consultant, CIOB
Dr Aeli Roberts MSc GDL BVC ICIOB	University College London
Dr Paul Sayer	Publisher, Wiley-Blackwell, John Wiley & Sons Ltd, Oxford

List of tables

List of figures

List of diagrams – Briefing Notes

0 Introduction

Project management

Project management has come a long way since its modern introduction to construction projects in the late 1950s. Now, it is an established discipline which executively manages the full development process, from the client's idea to funding coordination and acquirement of planning and statutory controls approval, sustainability, design delivery, through to the selection and procurement of the project team, construction, commissioning, handover, review, to facilities management coordination.

This *Code of Practice* positions the project manager as the client's representative, although the responsibilities may vary from project to project; consequently, project management may be defined as 'the overall planning, co-ordination and control of a project from inception to completion aimed at meeting a client's requirements in order to produce a functionally and financially viable project that will be completed safely, on time, within authorised cost and to the required quality standards'.

The fifth edition of this *Code of Practice* is the authoritative guide and reference to the principles and practice of project management in construction and development. It will be of value to clients, project management practices and educational establishments and students, and to the construction and development industries. Much of the information contained in the *Code of Practice* will also be relevant to project management practitioners operating in other commercial spheres.

Definitions

There are many definitions in existence for the term 'Project Management'. The CIOB, in this Code of Practice, and in all other publications, uses the following definition:

Project management

The overall planning, coordination and control of a project from inception to completion aimed at meeting a client's requirements in order to produce a functionally viable and sustainable project that will be completed safely, on time, within authorised cost and to the required quality standards.

Table 0.1 summarises a number of definitions of project management, as practiced by a selection of leading organisations involved in project management within the construction and building industry in UK.

Code of Practice for Project Management for Construction and Development, Fifth Edition. Chartered Institute of Building.
© Chartered Institute of Building 2014. Published 2014 by John Wiley & Sons, Ltd.

Table 0.1 Definitions of project management

Organisation	Definition of project management
Chartered Institute of Building	The overall planning, coordination and control of a project from inception to completion aimed at meeting a client's requirements in order to produce a functionally viable project that will be completed safely, on time, within authorised cost and to the required quality standards.
Association for Project Management	The application of processes, methods, knowledge, skills and experience to achieve the project objectives.[1]
British Standards 6079:2010	A unique set of coordinated activities, with definite starting and finishing points, undertaken by an individual or organisation to meet specific objectives within defined schedule, cost and performance parameters.
Office of Government Commerce (Department of Business, Innovation, and skills)	The planning, monitoring and control of all aspects of the project and the motivation of all those involved in it to achieve the project objectives on time and to the specified cost, quality and performance.[2]
International Organization for Standardization 21500:2012	Project management is the application of methods, tools, techniques and competencies to a project. Project management includes the integration of the various phases of the project lifecycle.
International Project Management Association[3] IPMA	Project management (PM) is the planning, organising, monitoring and controlling of all aspects of a project and the management and leadership of all involved to achieve the project objectives safely and within agreed criteria for time, cost, scope and performance/quality. It is the totality of coordination and leadership tasks, organisation, techniques and measures for a project. It is crucial to optimise the parameters of time, cost and risk with other requirements and to organise the project accordingly
Project Management Institute[4] PMI	Project management is the application of knowledge, skills and techniques to execute projects effectively and efficiently. It is a strategic competency for organisations, enabling them to tie project results to business goals – and thus, better compete in their markets.

[1]Definition as available at http://www.apm.org.uk/content/project-management (accessed November 2012).
[2]Definition obtained from OGC Glossary of Terms & Definitions v06 March 2008 – at the time of publication the document is available at www.gov.uk through publications of the Department of Business, Innovation & Skills.
[3]Definition obtained from ICB 3.0 – page 127.
[4]Definition as available at http://www.pmi.org/About-Us/About-Us-What-is-Project-Management.aspx (accessed February 2013).

Characteristics of construction projects

Construction projects have inherent features that make them highly complicated enterprises. These features are characterised by high levels of complexity, uncertainty and uniqueness and include

- Complexity created by the fragmentation of the organisational mechanism by which most projects are delivered. Usually the project delivery team is external to the client organisation, there is a separation between the designers and the constructors and the requirement for a wide range of specialist knowledge and skills demands the involvement of a large number of consultants, contractors, suppliers and statutory bodies.

- Complexity of the technology involved in the construction of modern buildings.

- Logistical complexity created by the locational aspects of projects – the site being a fixed location means that everything else must be taken to it. It is likely logistical complexity will be increased in a highly urbanised country where the pressure on land means the building footprint is likely to be the same as the site area, leaving minimal working space.

- Uncertainty created by exposure to the extremes of the weather.

- Uniqueness of each project; the project organisation and the participants vary, site conditions are different, technology adopted for the building varies, external influences on the project will be different and client constraints will be different.

- Uncertainty caused by the time necessary for the project life cycle. The longer the period of time, the greater the opportunity for the project to be impacted by changing external circumstances, such as economic conditions, or by changing client requirements.

Further pressures are created by a client needing to commit to key criteria such as the project duration and cost budget at an early stage, often before the full implications of what the project actually is about and how it is to be implemented have been developed in detail.

Most participants to the project are involved because they are offering a service or product as part of their business activity. It is usual practice for this involvement to be a formal contractual agreement with an agreed fixed, lump sum price based on a definition of the service or product required. Throughout their contribution to the project, participants are therefore balancing protecting their commercial position with working towards helping to achieve the overall project objectives. This relationship is not without difficulties and does not always work to the best advantage of the client or the project.

Characteristics of construction project management

Construction projects are intricate, resource consuming and often complex activities. The development and delivery of a project typically consists of several phases, sometimes over lapped but always linked, requiring a wide variety of skills and specialised services to balance the key project constraints (Figure 0.1). In progressing from initial feasibility to completion and occupation, a typical construction project passes through successive somewhat distinct stages that necessitate input from such asynchronous areas such as financial institutions, regulatory and statutory organisations, members of the public, engineers, planners, architects, specialist designers, cost engineers, building surveyors, lawyers, insurance companies, constructors, suppliers, tradesmen and cost managers.

During the construction stage itself, a project of relatively simple design and methodology involves a wide range of skills, materials and a plethora of different but often sequential activities and tasks that must follow a predetermined order that constitutes a complicated and sensitive pattern of individual criteria and restrictive sequential relationships.

The Construction Industry Council (CIC) suggests that the primary purpose of project management is to add significant and specific value to the process of delivering construction projects.[1] This is achieved by the systematic application of a set of generic project-orientated management principles throughout the life of a project. Some of these techniques have been tailored to the sector requirements unique to the construction industry.

The function of project management is applicable to all projects. However, on smaller or less complex projects, the role may well be combined with another discipline, for example, leader of the design team. The value added to the project by project management is unique: no other process or method can add similar value, either qualitatively or quantitatively.

---

[1] Construction project management skills.pdf, at http://www.cic.org.uk (accessed April 2014).

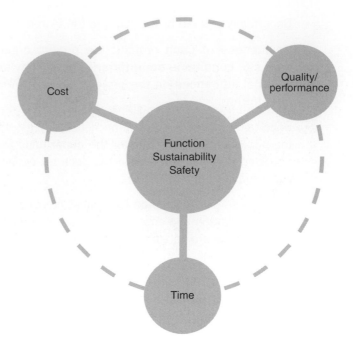

Figure 0.1 Key project constraints.

Adding value

The raising of standards should lead significantly to the adding of value. Greater awareness can result in better design, improved methods and processes, new material choices, less waste, decreases in transportation costs and ultimately more efficient buildings, all of which can bring added value to the whole development process.

Scope of project management

Construction and development projects involve the coordinated actions of many different professionals and specialists to achieve defined objectives. The task of project management is to bring the professionals and specialists into the project team at the right time to enable them to make their best possible contribution, efficiently.

Professionals and specialists bring knowledge and experience that contributes to decisions, which are embodied in the project information. The different bodies of knowledge and experience all have the potential to make important contributions to decisions at every stage of projects. In construction and development projects, there are far too many professionals and specialists involved for it to be practical to bring them all together at every stage. This creates a dilemma because ignoring key bodies of knowledge and experience at any stage may lead to major problems and additional costs for everyone.

The practical way to resolve this dilemma is to carefully structure the way the professionals and specialists bring their knowledge and experience into the project team. The most effective general structure is formed by the eight project stages used in this *Code of Practice*'s description of project management.

Project lifecycle

The different stages of the project lifecycle as identified across the industry have been summarised and compared in Figure 0.2.

Code of Practice for Project Management	Royal Institute of British Architects (RIBA) Plan of Work 2013	BIM Digital Plan of Work 2013	BS 6079-1:2010	ISO 21500:2012
1 Inception	0 Strategic definition	1 Strategy	1 Conception	1 Initiating
2 Feasibility	1 Preparation and brief	2 Brief	2 Feasibility	2 Planning
3 Strategy	2 Concept design	3 Concept		
		4 Definition		
4 Pre-construction	3 Developed design	5 Design	3 Realisation	3 Implementing
	4 Technical design			
5 Construction	5 Construction	6 Build & commission		
6 Testing and commissioning				4 Controlling
7 Completion, handover and operation	6 Handover & close out	7 Handover & close out	4 Operation	
8 Post-completion review and in use	7 In use	8 Operation & end of life	5 Termination	5 Closing

Figure 0.2 Project lifecycle.

In many projects, there will be a body of knowledge and experience in the client organisation which has to be tapped into at the right time and combined with the professional and specialists' expertise.

Each stage in the project process is dominated by the broad body of knowledge and experience that is reflected in the stage name. As described earlier, essential features of that knowledge and experience need to be taken into account in earlier stages if the best overall outcome should be achieved. The way the professionals and specialists who own that knowledge and experience are brought into the project team at these earlier stages is one issue that needs to be decided during the strategy stage.

The results of each stage influence later stages, and it may be necessary to involve the professionals and specialists who undertook earlier stages to explain or review their decisions. Again, the way the professionals and specialists are employed should be decided in principle during the strategy stage.

Each stage relates to specific key decisions (see Table 0.2) Consequently, many project teams hold a key decision meeting at the end of each stage to confirm that the necessary actions and decisions have been taken and the project can therefore begin the next stage. There is a virtue in producing a consolidated document at the end of each stage that is approved by the client before proceeding to the next stage. This acts as a reference mark as well as acting as a vehicle for widespread ownership of the steps that have been taken.

Having considered the social, economic and environmental issues, projects begin with the inception stage which starts with the business decisions by the client that suggest a new construction or development project may be required. Essentially, the inception stage consists of commissioning a project manager to undertake the next stage which is to test the feasibility of the project. The feasibility stage is a crucial stage in which all kinds of professionals and specialists may be required to bring many kinds of knowledge and experience into a broad ranging evaluation of feasibility. It establishes the broad objectives and an approach to sustainability for the project, and so exerts an influence throughout subsequent stages.

The next stage is the strategy stage which begins when the project manager is commissioned to lead the project team to undertake the project. This stage requires the project's objectives, an overall strategy and procedures in place to manage the sustainability and environmental issues, and the selection of key team members to be considered in a highly interactive manner. It draws on many different bodies of knowledge and experience and is crucial in determining the success of the project. In addition to selecting an overall strategy and key team members to achieve the project's objectives, it determines the overall procurement approach and sets up the control systems that guide the project through to the final post-completion review and project close-out report stage. In particular, the strategy stage establishes the objectives for the control systems. These deal with much more than quality, time and cost. They provide agreed means of controlling value from the client's point of view, monitoring time and financial models that influence the project's success, managing risk, making decisions, holding meetings, maintaining the project's information systems and all the other control systems necessary for the project to be undertaken efficiently.

At the completion of the strategy stage, everything is in place for the pre-construction stage. This is when the design is developed and the principal decisions are made concerning time, quality and cost management. This stage also includes statutory approvals and consents, considering utility provisions such as water and electricity, monitoring of the environmental performance targets, and bringing manufacturers, contractors and their supply chains into the project team. Like the earlier stages, the

Table 0.2 Specific key decisions

Project stages	Key high-level processes	Key high-level objective	Key high-level deliverables	Key high-level resources (key driver)
Stage 1: Inception	Project need Project manager selection (optional) Project mandate Environmental mandate	'What is the need?'	Project initiation document (PID)	Client team Project manager
Stage 2: Feasibility	Project brief Project manager selection Feasibility studies Business case Funding options Delivery parameters	'Is the need feasible?'	Project brief Signing off business case	Client team Project manager Specialist consultants
Stage 3: Strategy	Project governance Parameters Project strategy Project organisation and control Accountability and responsibility Procurement strategy Selection and appointment of project team Tender procedure Project execution plan	'How will the need be realised?'	Project execution plan	Client team Project manager Specialist consultants
Stage 4: Pre-construction	Design delivery process Technical design and production information Value management Procurement of supply chain Contractual arrangements	'What do we need to build? How would it look like and function? How would we deliver it and manage it?'	Design outputs Contractual arrangements	Client team Project manager Design team CDM coordinator

(Continued)

Table 0.2 (Continued)

Project stages	Key high-level processes	Key high-level objective	Key high-level deliverables	Key high-level resources (key driver)
Stage 5: Construction	Performance monitoring and control Health, safety and welfare systems Quality management and control	'Are we constructing what has been designed?'	Performance management plan	Client team Project manager Design team CDM coordinator Constructor team
Stage 6: Testing and commissioning	Commissioning services Commissioning documentation	'Is the building working as designed?'	Commissioning documentation	Client team Project manager Design team CDM coordinator Constructor team Commissioning team
Stage 7: Completion, hand over and operation	Planning and scheduling handover Handover procedures Operational commissioning Client occupation	'How do we use the building?'	Handover documentation Health and safety file	Client team Project manager Design team CDM coordinator Construction team Commissioning team Occupation and maintenance team
Stage 8: Post-completion review and in use	Post-occupancy evaluation Project audit Project feedback Close-out report Benefits realisation	'Has the project satisfied the need?'	Project close-out report Post-occupancy Evaluation Occupation strategy	Client team Project manager Occupation and maintenance team

pre-construction stage often requires many different professionals and specialists working in creative and highly interactive ways. It is therefore important that this stage is carefully managed using the control systems established during the strategy stage to provide everyone involved with relevant, timely and accurate feedback about their decisions. Completion of this stage provides all the information needed for construction to begin.

The construction stage is when the actual building or other facility that the client needs is produced. In modern practice, this is a rapid and efficient assembly process delivering high-quality facilities. It makes considerable demands on the control systems, especially those concerned with time and quality. The complex nature of modern buildings and other facilities and their unique interaction with a specific site means that problems will arise and have to be resolved rapidly. Information systems are tested to the full, design changes have to be managed, construction and fitting out teams have to be brought into the team and empowered to work efficiently. Costs and time have to be controlled within the parameters of project objectives and the product delivered to the quality and specification as set previously.

The construction stage leads seamlessly into a key stage in modern construction and development projects: the commissioning stage. The complexity and sophistication of modern engineering services makes it essential that time is set aside to test and fine-tune each system. Any environmental performance targets such as Building Research Establishment Environmental Assessment Method (BREEAM) certification can be used as a measure of the project's performance. Therefore, these activities form a distinct and separate stage which should predominantly be complete before beginning the completion, handover and operation stage which is when the client takes over the practically completed building or other facility. In some instances, there may also be some post-occupation commissioning and testing.

The client's occupational commissioning needs to be managed as carefully as all the other stages because it can have a decisive influence on the project's overall success and environmental performance. New users always have much to learn about what a new building or other facility provides. They need training and help in making best use of their new building or other facility. It is good practice for their interests and concerns to be considered during the earlier stages and preparation for their move into the new facility at the right time so that there are no surprises when the client's organisation takes occupation.

The final stage is the post-completion review and in-use stage. This provides the opportunity for the project team to consider how well the project's objectives have been met and what lessons should be taken from the project. A formal report describing these matters provides a potentially important contribution to knowledge. For clients who have regular programmes of projects and for project teams that stay together over several projects, such reports provide directly relevant feedback. Even where this is not the case, everyone involved in a project team, including the client, is likely to learn from looking back at their joint performance in a careful objective review. Projects where a BIM protocol had been established, then information exchange between the delivery team and the operations team will form a key highlight of this stage. In some projects, the client may wish to extend the services of the project manager (and may be the BIM manager) to facilitate the transition from delivery to operation, including assessment of project benefits and updating the controls and procedures as necessary.

1 Inception

Stage checklist

Key processes:	Project need
	Project manager selection
	Project mandate
	Environmental mandate
	BIM mandate
Key objective:	'What is the need?'
Key deliverables:	Project mandate (project initiation document)
Key resources:	Client team
	Project manager

Stage process and outcomes

Inception is the initial stage of the development process; it is a transition between the client's strategic business decision making and the implementation of a project. The stage confirms a need, either business or social, that requires some form of capital development and concludes with the client making a decision to proceed with a detailed appraisal of the viability of the development.

Principally this is a client-led process, but depending on the nature of the client and the complexity of determining the client's requirements it may involve the services of management consultants or a professional adviser and these may be in-house or external to the client's organisation.

Outcomes:

- Statement of the key business objectives, project mandate and constraints
- Statement of an environmental mandate
- Outline of BIM strategy
- Definition of the project management structure
- Approval to proceed to the feasibility stage
- Appointment of the project manager

Code of Practice for Project Management for Construction and Development, Fifth Edition. Chartered Institute of Building.
© Chartered Institute of Building 2014. Published 2014 by John Wiley & Sons, Ltd.

The client

Client obligations and responsibilities

The client organisation will need to ascertain what the needs and objectives are that the project is aiming to satisfy and how the project fits in with their strategic objectives.

The client organisation will also need to establish that it has the resources to develop and deliver the project, including articulating 'vision' and the 'need' into tangible strategies and objectives as well as understanding and delivering its responsibilities and obligations as a client. Having determined the degree of their involvement in the development of the project the client will need to review the extent of external support required.

Client project objectives

The main objective at this stage for the client is to make the decision to invest in a construction or development project. The client should have prepared a project mandate (capital expenditure programme) which will evolve into a business case for the project involving careful analysis of its business, organisation, present facilities and future needs. Experienced clients may have the necessary expertise to prepare their project mandate themselves. Less experienced clients may need help. Many project managers are able to contribute to this process. This process will result in a project-specific statement of need. The client's objective will be to obtain a totally functional facility, which satisfies this need and must not be confused with the project objectives, which will be developed later from the statement of need.

A sound project mandate will:

- be driven by needs
- be based on sound information and reasonable estimation
- contain rational processes
- be aware of the risks associated
- contain flexibility
- maximise the scope of obtaining best value from resources
- utilise previous experience
- incorporate sustainability cost-effectively

Client engagement: Internal team

Investment decision maker: This is typically a corporate team of senior managers and/or directors who review the potential project and monitors the progress. However, the team seldom is involved directly in the project process.

Project sponsor: Typically a senior person in the client's organisation, acting as the focal point for key decisions about progress and variations. The project sponsor has to possess the skills to lead and manage the client role, have the authority to take day-to-day decisions and have access to people who are making key decisions.

Client's advisor: The project sponsor can appoint an independent client advisor (also referred to as construction advisor or project advisor or independent client advisor) who will provide professional advice in determining the necessity of construction and means or procurement, if necessary. If advice is taken from a consultant or a contractor, those organisations have a vested interest not only in confirming the client's need, but also in selling their services and products.

The client advisor can assist with:

- project mandate and business case development (see 'Feasibility' stage)

- investment appraisal

- designing and planning for sustainability

- understanding the need for a project

- deciding the type of project that meets the need

- generating and appraising options (when appropriate)

- selecting an appropriate option (when available)

- risk assessment (when appropriate)

- advising the client on the choice of procurement route

- selecting and appointing the project team

- measuring and monitoring performance (when appropriate)

The client advisor should understand the objectives and requirements of the client but should remain independent and objective in providing advice directly to the client. Other areas where the client may have sought independent advice include chartered accounting, tax and legal aspects, market research, town planning, chartered surveying and investment banking.

Project manager

Project managers can come from a variety of backgrounds, but all will need to have the necessary skills and competencies to manage all aspects of a project from inception to occupation. This role may be fulfilled by a member of the client's organisation or be an external appointment.

Project manager's objectives

The project manager, both acting on behalf of, and representing the client, has the duty of 'providing a cost-effective and independent service, selecting, correlating, integrating and managing different disciplines and expertise, to satisfy the objectives and provisions of the project brief from inception to completion. The service provided must be to the client's satisfaction, safeguard his interests at all times, and, where possible, give consideration to the needs of the eventual user of the facility'.

The key role of the project manager is to motivate, manage, coordinate and maintain the morale of the whole project team. This leadership function is essentially about managing people and its importance cannot be overstated. A familiarity with all the other tools and techniques of project management will not compensate for shortcomings in this vital area. Further guidance on the leadership aspect of the project manager's role has been provided in Briefing Note 1.01 at the end of this section.

In dealing with the project team, the project manager has an obligation to recognise and respect the professional codes of the other disciplines and, in particular, the responsibilities of all disciplines to society, the environment and each other. There are differences in the levels of responsibility, authority and job title of the individual responsible for the project, and the terms project manager, project coordinator and project administrator are all widely used.

It is essential, in order to ensure an effective and cost-effective service, that the project should be under the direction and control of a competent practitioner with a proven project management track record developed from a construction industry-related professional discipline. This person is designated the project manager and is to be appointed by the client with full responsibility for the project. Having delegated powers at inception, the project manager may exercise, in the closest association with the project team, an executive role throughout the project with appropriate input from the client.

Project manager's duties

The duties of a project manager will vary depending on the client's expertise and requirements, the nature of the project, the timing of the appointment and similar factors. If the client is inexperienced in construction, the project manager may be required to develop his own brief. Whatever the project manager's specific duties in relation to the various stages of a project, there is the continuous duty of exercising control of project time, cost and performance. Such control is achieved through forward thinking and the provision of good information as the basis for decisions for both the project manager and the client. A matrix correlating suggested project management duties and client's requirements is given in Table 1.1.

An example of typical terms of engagement for a project manager is outlined in Briefing Note 1.02. It will be subject to modifications to reflect the client's objectives, the nature of the project and contractual requirements.

The term 'project coordinator' is applied where the responsibility and authority embrace only part of the project, for example, pre-construction, construction and handover/migration stages. (For professional indemnity insurance purposes a distinction is made between project management and project coordination. When the project manager appoints other consultants the service is defined as project management and when the client appoints other consultants the service is defined as project coordination.)

Project manager's appointment

It is advisable to appoint the project manager at the inception stage so that the project manager can advise and become involved in the option appraisal process. This should ensure professional, competent management coordination, monitoring and controlling of the project to its satisfactory completion, in accordance with the client's brief. However, depending on the nature and type of the project and the client's in-house expertise, the project manager could be appointed as late as the start of the strategy stage, but this could deprive them of important background information and is therefore not generally recommended.

Project mandate

The project mandate could be defined as the authority given to the project team to develop and progress the project within given and agreed boundaries, set by the client.

These will include requirements on programme for delivery of the project, the budget and also the requirements for the finished building in terms of function, quality and any particular requirements on performance, such as environmental performances.

Understanding the client need as clearly as possible at the start of the project is fundamental to project success.

The project mandate (also referred to as initial project inquiry (IPI) or project initiation document (PID)) is usually the first document produced to trigger a project. It is not seen as a project documentation but as a pre-project document. However, often the trigger to a project is poor and it is advisable to put together a document, which encapsulated the

Table 1.1 Duties of project manager

Duties*	Client's requirements			
	In-house project management		Independent project management	
	Project management	Project coordination	Project management	Project coordination
Be named in the contract	■		+	
Assist in preparing the project brief	■		■	
Develop project manager's brief	■		■	
Advise on budget/funding/programme/ risk management arrangements	■		+	
Advise on site acquisition, grants and planning	■		■	
Arrange feasibility study and report	■	+	■	+
Develop project strategy	■	+	■	+
Prepare project handbook	■	+	■	+
Develop consultant's briefs	■	+	■	+
Devise project programme	■	+	+	+
Select project team members	■	+	■	+
Establish management structure	■	+	■	+
Coordinate design processes	■	+	■	+
Appoint consultants	■	■	■	+
Arrange insurance and warranties	■	■	■	+
Select procurement system	■	■	■	+
Arrange tender documentation	■	■	■	+
Organise contractor prequalification	■	■	■	+
Evaluate tenders	■	■	■	+
Participate in contractor selection	■	■	■	+
Participate in contractor appointment	■	■	■	+
Organise control systems including reporting procedures	■	■	■	■
Monitor progress	■	■	■	■
Manage and monitor meetings	■	■	■	■
Authorise payments	■	■	■	+
Organise communication/reporting systems	■	■	■	■
Provide project coordination	■	■	■	■
Issue health and safety procedures	■	■	■	■
Address environmental aspects	■	■	■	■
Coordinate statutory authorities	■	■	■	■
Monitor budget and changes	■	■	■	■
Develop final account	■	■	■	■
Arrange pre-commissioning/ commissioning	■	■	■	■

(Continued)

Table 1.1 (Continued)

Duties*	Client's requirements			
	In-house project management		Independent project management	
	Project management	Project coordination	Project management	Project coordination
Organise handover/occupation	■	■	■	■
Advise on marketing/disposal	■	+	■	+
Organise maintenance manuals	■	■	■	+
Plan for maintenance period	■	■	■	■
Develop maintenance programme/ staff training	■	■	■	+
Plan facilities management and coordinate BIM	■	■	■	+
Arrange for feedback monitoring and post-completion review	■	■	■	+
Investigate BIM implementation	■	■	■	■
Liaise with funding institutions	■	■	■	+
Liaise with ground landlord	■	+	■	+
Liaise on acquisition, valuation, disposal of land	■	+	■	+
Liaise with agents over leasing tenants queries, etc.	■	+	■	+
Liaise with client over move to new premises	■	+	■	+
Liaise coordination with legal agents	■	+	■	+
Advise and manage client's changes	■	■	■	■

Symbols: ■, suggested duties; +, possible additional duties.
*Duties vary by project, and relevant responsibility and authority.

ideas and any basic information that can be identified at this point. Some key questions that may be considered while preparing the project mandate include:

- Is the level of authority commensurate with the anticipated size, risk and cost of the project?
- Is there sufficient detail to allow the appointment of key team members including the project manager?
- Are all the known (internal) stakeholders identified?
- Does the project mandate identify what is necessary for the project to be a success (key success criteria)?

An indicative template for project mandate is outlined in Briefing Note 1.03.

Environmental mandate

Environmental performance and impact may be particularly important to the client. Corporate Social Responsibility plays an important role in the delivery of built environment projects.

Environmental mandate includes requirements for the environmental performance of the building. It may also include requirements on carbon emissions and energy consumption.

In addition, it may also prescribe requirements for environmental impact on the local topography or adjacent area. Lastly it may determine outcomes in terms of the local community, such as providing employment and training opportunities or the use of local supply chain.

An environmental mandate for the project will provide the management framework for the planning and implementation of construction activities in accordance with the environmental commitments of the organisation, the project context, funders, project end users or any other stakeholders.

The environmental mandate will influence key design parameters relating to sustainability, performance and operational technologies.

The environmental mandate should also outline the overall environmental management criteria for the project including what are the key success factors for the projects in terms of environmental management.

BIM mandate

BIM (building information modelling) enables the sharing of information and data between all stakeholders and participants around the whole asset lifecycle. It provides a platform for consistent, structured, perfect data, to enable informed smart decision making at all stages of the project process.

If BIM is to be used on a project, then this should be implemented right from the start. As industry adopts BIM as the normal way of working, this will become standard practice. However for the moment, migrating the project to BIM might occur at any stage. Naturally this has consequences in terms of cost, time, resources required and scale of difficulty.

When BIM is being used, then it is important to establish the drivers for this. If it is client driven, what does the client require of the project BIM?

Is it simply for efficiency of process or will outputs be required at various stages in accordance with COBie (Construction Operations Building Information Exchange) advise, and at handover the model and data sets used for FM/operations and integration with their building management systems?

A Project BIM Execution Protocol (see BIM Protocol – Standard Protocol for use in projects using Building Information Models – CIC/BIM Pro – first edition 2013) must be established to ensure BIM is used to maximum advantage and that the whole team is working together in a consistent manner.

Briefing Note 1.01 Leadership in project management

What is leadership?

Leadership, as a management attribute, has been subjected to a significant amount of attention. Defining simplistically, 'it is the process in which an individual influences other group members towards the attainment of group or organisational goals' (Shackleton[1]). Inevitably, there are a wide range of theories and schools of thoughts encompassing this subject (a number of reference documents are listed in the bibliography). The latest discussions tend to focus on transactional and transformational natures of leadership.

Leadership and project manager

The very definition of construction project management implies that within a defined timescale, the project is expected to achieve an agreed set of targets utilising specific resources. This requires not only a very efficient project manager but also an effective project leader who can lead the project team spontaneously, mainly focusing towards the project and motivating the project team members to achieve the targets within the agreed project framework. The key aspects or traits that a successful project manager would require to excel in are motivation, performance appraisal, resource allocation and management, and planning and communication.

What are the traits of effective leaders?

There are a range of theories outlining leadership styles and traits. Broadly, successful leadership traits are characterised in six styles as detailed in the following table.

Leadership styles

Style	Result
Coercive	Leader demands immediate compliance
Authoritative	Leader mobilises people towards a vision
Affiliative	Leader creates emotional bonds and harmony
Democratic	Leaders use participation to create consensus
Pace-setting	Leader expects excellence and self direction from the group
Coaching	Leader develops people for the future

[1] Shackleton, V. (1995) *Business Leadership*. Routledge, London.

The general suggestion[2] is that the leaders need to understand how the styles relate to their individual competencies and situational requirements so as to identify the most suitable approach.

There are some differences of views on effectiveness of training of leadership skills (are leaders born or made?). However, it is advisable to stress the need for flexibility of the leader – to learn to lead differently depending on the situational and contextual needs; hence, the leaders should learn many styles and learn to diagnose the needs of the context and situation.

Are there any quick wins?

Although different styles and tactics would suit different contexts and situations, the adaptation of the following should enhance effective project management:

- acknowledging positive contribution
- ensuring open communication
- 'touching base' with the team members on a regular basis
- sharing praise

---

[2] Goleman, D. (2000) Leadership that gets results. *Harvard Business Review*, March–April.

Job title: Project manager.

Date effective:

General objective

Acting as the client's representative within the contractual terms applicable, to lead, direct, coordinate and supervise the project in association with the project team.

The project manager will ensure that the client's brief, all designs, specifications and relevant information are made available to, and are executed as specified with due regard to cost by, the design team, consultants and contractors (i.e. the project team) so that the client's objectives are fully met.

Relationships

Responsible and reporting to   The client.

Subordinates   Practice support staff and secretarial/clerical staff.

Functional   Fully integrated working with any project support staff who are not line subordinates:

1. liaison, as required/expedient with relevant client's staff, for example, legal, insurance, taxation

2. full interdependent cooperation with

   (a) design team and consultants

   (b) contractors

   (c) client and other key stakeholders

External   Liaison with local or other relevant authority on matters concerning the project. Contact with suppliers of construction materials/equipment, in order to be aware of the most efficient and cost-effective application, and working methods.

Contact with

1. Client's information and communication technology (ICT) team or other higher technology sources, able to provide expertise on the application of advanced technology in the design and/or construction processes of the project (e.g. communications, environment, security and fire prevention/protection systems).

2. And preferably, membership of appropriate professional bodies/societies.

(Continued)

Authority

The definition of the authority of the project manager is a key requirement in enabling him to manage the successful achievement of the client's objectives. The extent must be clearly defined. A distinction should be drawn between the responsibility that the project manager may have which concerns his accountability for different aspects of the project and the authority which will determine the ability of the project manager to control, command and determine the commitment of resources to the project. The full extent of the responsibility and authority vested in the project manager will depend on the terms and duties included in the project management agreement.

The extent of the project manager's responsibility and authority may be balanced, but the two may be unequal. Frequently, the project manager may have extensive responsibility in an area that does not carry commensurate authority, or vice versa.

The authority of the project manager should be defined regarding his obligations to issue instructions, approve limits of expenditure and when to notify the client and seek the instructions of the client in matters relating to

1. the schedule and time taken to complete the project

2. expenditure and costs, including development budget, project cost plan, and financial rewards and viability

3. designs, specifications and quality

4. function

5. contractors' contracts

6. consultants' appointments

7. assignment of contracts or appointments

8. administrative procedures, including issuing or signing of correspondence, certification and other project documentation

The client and the project manager should give careful consideration to the authority that will be necessary to ensure the successful achievement of the client's objectives and, if necessary, establish appropriate lines of authority and communication within the client organisation to facilitate the implementation of agreed procedures.

Detailed responsibilities and duties

1. Analysis of the client's objectives and requirements, assessment of their feasibility and assistance in the completion of project brief and establishment of the capital budget.

2. Formulation, for the client's approval, of the strategic plan for achieving the stated objectives within the budget, including, where applicable, the quality assurance scheme.

3. Generally keeping the client informed, throughout the project, on progress and problems, design/budgeting/construction variations and such other matters considered to be relevant.

4. Participation in making recommendations to the client, if required, in the following areas:

   • The selection of the consultants as well as in the negotiation of their terms and conditions of engagement.

- The appointment of contractors/subcontractors, including the giving of advice on the most suitable forms of tender and contract.

5. Preparation for the client's approval of the following items:

    - The overall project schedule embracing site acquisition, relevant investigations, planning, pre-design, design, construction and handover/occupation stages.

    - Proposals for architectural and engineering services. The project manager will monitor progress and initiate appropriate action on all submissions concerned with planning approvals and statutory requirements (timely submission, alternative proposals and necessary waivers).

    - The project budget and relevant cash flows, giving due consideration to matters likely to affect the viability of the project development.

6. Finalisation of the client's brief and its confirmation to the consultants. Providing the client with all existing and, if necessary, any supplementary data on surveys, site investigations, adjoining owners, adverse rights or restrictions and site accessibility/traffic constraints.

7. Recommending to the client and securing approval for any modifications or variations to the agreed brief, approved designs, schedules and/or budgets resulting from discussions and reviews involving the design team and other consultants.

8. Setting up the management and administrative structure for the project and thereby defining:

    - responsibilities and duties, as well as lines of reporting, for all parties

    - procedures for clear and efficient communication

    - systems and procedures for issuing instructions, drawings, certificates, schedules and valuations and the preparation and submission of reports and relevant documentary returns

9. Agreeing tendering strategy with the consultants concerned.

10. Advising the client as necessary on the following items:

    - The progress of the design and the production of required drawings/information and tender documents, stressing at all times the need for a cost-effective approach to optimise costs in construction methods, subsequent maintenance requirements, preparation of tender documents and performance/workmanship warranties.

    - The correctness of tender documents.

    - The prospective tenderers pre-qualified by the design team and other consultants involved, obtaining additional information if pertinent and confirming accepted tenders to the client and the consultants.

    - The preliminary construction schedule for the main contractors, agreeing any revisions to meet fully the client's requirements and releasing this to the project team for action.

    - The progress of all elements of the project, especially adherence to the agreed capital and sectional budgets, as well as meeting the set standards and initiating any remedial action.

    - The contractual activities the client must undertake, including user study groups and approval/decision points.

1.02 Briefing Note

11. Establishing with the quantity surveyor the cost monitoring and reporting system and providing feedback to the other consultants and the client on budget status and cash flow.

12. Organising and/or participating in the following activities:

    - Presentations to the client, with advice on and securing approval for the design of fabrics, finishes, fitting-out work and the environment of major interior spaces.

    - All meetings with the project team and others involved in the project (chairing or acting as secretary) to ensure

        - an adequate supply of information/data to all concerned

        - that progress is in accordance with the schedule

        - that costs are within the budgets

        - that required standards and specifications are achieved

        - that contractors have adequate resources for the management, supervision and quality control of the project

        - that the relevant members of the project team inspect and supervise construction stages as specified by the contracts

13. Responsible for:

    - preparation of the project handbook

    - achieving good communications and motivating the project team

    - monitoring progress, costs and quality and initiating action to rectify any deviations

    - setting priorities and effective management of time

    - coordinating the project team's activities and output

    - monitoring project resources against planned levels and initiating necessary remedial action

    - preparing and presenting specified reports to the client

    - submitting time sheets and other data on costing and control to the client

    - processes, including required returns and all other relevant information

    - approving, in collaboration with the project team and within the building contract provisions, any sublet work

    - identifying any existing or potential problems, disputes or conflicts and resolving them, with the cooperation of all concerned in the best interests of the client

    - recommending to the client the consultants' interim payment applications and monitoring such applications from contractors

    - monitoring all pre-commissioning checks and progress of any remedial defects liability work and the release of retention monies

    - verifying with the project team members concerned any claims for extensions of time or additional payments and advising the client accordingly

    - checking consultants' final accounts before payment to the client

    - monitoring the preparation of contractors' final accounts, obtaining relevant certificates and submitting them for settlement by the client

- ensuring the inclusion in the contract and subsequently requesting the design team, consultants and contractors to supply the client with as-built and installed drawings, operating and maintenance manuals, and health and safety file, as well as ensuring arrangements are made for effective training of the client's engineering and maintenance staff, that is, facilities management

14. Taking all appropriate steps to ensure that site contractors and other regular or casual workers observe all the rules, regulations and practices of safety and fire prevention/protection. Exercising 'good site housekeeping' at all times.

15. Participating in the final cost reconciliation or final account of the project and taking such action as directed or required.

Extra-project activities

Participating in informal discussions with own and other practices, as well as the client's staff, on technical details, methods of operations, problem solving and any other pertinent actions relevant to present or previous projects, in order to exchange views/knowledge conducive to providing a more effective overall performance.

The project manager has responsibility for the following areas:

1. Personnel matters relating to his staff, including appraisal/reviews, training/ development and job coaching and counselling, as defined by the client and/or project management practice guidelines and procedures.

2. Updating himself and staff in new ideas relating to project management, including management/supervisory skills and practice generally, business, financial, legal and economic trends, the latest forms of contract, planning and Building Regulations, as well as advances in construction techniques, plant and equipment.

Terms of engagement: The services contracts

1. The CIC Consultants Contract Conditions and Scope of Services 2007 (2nd ed. 2011)

2. RICS Project Management Agreement (3rd ed. 1999)

3. APM Terms of Appointment for a Project Manager (1998)

4. NEC3 Professional Services Contract (PSC) (2005)

5. RIBA Form of Appointment for Project Managers (2010–2012 edition)

1.02 Briefing Note

1. Document and distribution history

2. Purpose: The purpose of this document is to define the project, to form a firm basis for management and the assessment of overall success. Information included in this document will ideally answer the following questions:

   - What is the project aiming to achieve?

   - Why it is important to achieve it?

   - Who will be involved and what are their responsibilities?

   - How and when will it happen?

   - How and what needs to be communicated?

   It is essential to obtain a clear view of the final objectives and outcome, as well as constraints and assumptions, that impact on those responsible for the project. This document outlines all of the essentials to form a firm foundation for defining the project, including objectives, outcomes, expectations, scope and time frames.

3. Authority responsible

4. Project background and context

5. Project objectives

6. Financial objective

7. Project outcomes and milestones – what is to be delivered and when

8. Project organisation structure – project management, project team and project board

9. Communications plan – inter- and intra-organisation

10. Initial project plan – activity schedule with resources indicating ownership

11. Project controls – who reports what to whom, when and how

12. Risk analysis – with contingency sums where necessary and or avoid transfer mitigate share accept

13. Project success criteria

14. Associated documents

15. Items out of scope

16. Any other relevant information

Briefing Note 1.04   Project handbook

Introduction

The purpose of the handbook is to guide the project team in the performance of its duties, which are the design, construction and completion of a project to the required specifications within the approved parameters of the contract budget and to schedule. In practice, a project handbook should be concise, clear and consistent with all other contract documentation and terms of engagement. The emphasis should be to identify policies, strategies and the lines of communication and key interfaces between the various parties. It is important that the handbook is tailored to fit the needs of each project. The comprehensive format given here would be too bulky for some projects with the danger of it being ignored.

The handbook is prepared by the project manager in consultation with the project team where possible at the beginning of the pre-construction stage and describes the general procedures to be adopted by the client and the team. It comprises a set of ground rules for the project team. It differs from the project execution plan, which is primarily written for the client and funding partners, in giving a route map through the stages and processes of the project demonstrating financial control and a *modus operandum* to achieve the project objectives.

The handbook is not a static document and it is anticipated that changes and amendments will be required in accordance with procedures as later outlined. Consequently, a loose-leaf format should be adopted to facilitate its updating by the project manager who is the only person authorised to coordinate and implement revisions. Copies of the handbook will be provided to each nominated member of the project team as listed under parties to the project.

Aims of the handbook

The aim of the handbook is to identify responsibilities and coordinate the various actions and procedures from other documents/data already or currently or likely to be prepared into one authoritative document covering as a rule and depending on the nature/scope of project, the main elements and activities outlined in the following sections.

Parties to the project

This section will include the following items:

- A list of all parties involved in the project including those employed by the client as well as their contact details (addresses, phone and fax numbers and email address).

- The name of the project manager responsible for the project together with details of his duties responsibilities and authority.

- Details of other team members and/or stakeholders involved complete with their duties, responsibilities and contact details.

- Organisational charts indicating line and functional relationships, contractual and communication links and any changes to suit the various stages/phases of the project.

Third parties

This section will provide the names and contact details of all legal authority departments, public utilities, hospitals, doctors, police stations, fire brigade, trade associations, adjoining landowners, adjacent tenants and any other bodies or persons likely to be involved.

Roles and duties of the project team

The information provided should be the minimum necessary to facilitate the understanding of the roles of the others involved by each member of the team. The services to be provided are described by reference to standard agreements or contracts with any amendments and additions included. The aim is to ensure that there are no gaps or overlaps.

Project site

Details will be provided of prevailing relevance of arrangements for demolition, clearing and diversion of existing services, hoardings and protection to adjacent areas (e.g. noise pollution).

General administration including communication and document control

The project manager will be responsible for the following items:

- The adequacy of all aspects of project resourcing (staffing, equipment and aids, site offices and welfare accommodation).

- Office operating systems and routines so that the staff know them and are applied consistently and efficiently.

- Providing suitable working accommodation and facilities for members of the project team and for meetings or group discussions.

Action will need to be taken by the project manager in respect of documentation control, storage, location and retrieval; this will affect:

- letters, contract documents, reports, drawings, specifications, schedules, including financial and all specialist fields (e.g. facilities management, technology, health and safety, environmental)

- accessibility for updating

- records for all documents/files and control of their movement

- office security: (1) storage of legal documents (originals and duplicates); (2) entry safeguard, fire and intruder alarms

- retention of documents/files on project completion/suspension: (1) archive storage – legal and contractual time limits; (2) dead files – removals and destruction and their register

All correspondence should be headed by the project title and identified by:

- subject/reference of communication

- addressee's full details

- those parties receiving copies

Each piece of correspondence should refer to a single matter or a series of direct and closely linked matters only. Distribution of copies should be decided on the basis of the subject matter and confidentiality against a predetermined list of recipients. All communications between the parties of the project involving instructions must be given in writing and the recipients should also confirm it in the same manner.

Contract administration

Contract conditions

It is essential that there is an understanding of the terms of all contracts and their interpretation by all concerned. The role of parties, their contribution and responsibilities, including relevance of timescales and client–project manager operating and approvals pattern, will have to be established.

Contract management and procedures

Matters associated with contract management will include forms of contract for contractors/subcontractors; works carried out under separate direct contract; procedures for the selection and appointment of contractors/subcontractors; checklists for design team members and consultants meeting their supervisory and contractual obligations (e.g. inspections and certification); the placement of orders for long delivery components and the preparation of contract documents.

Tender documentation

Design and specifications details to be included; tender analysis and reporting; lists of tenderers and interview procedures; system for the preparation of documents and their checking; award and signature arrangements.

Assessment and management of variations

Extensions of time
- The project manager has the responsibility for ensuring that there is early warning, hence creating the possibility of alternative action/methods to prevent delay and additional costs.

- A schedule should be prepared, stating the grounds for extension, relevant contract clauses and forecast of likely delay and cost.

- The involvement and possible contribution to the solution of problems of other parties affected should be established.

- A procedure will need to be available for extension approval. If relevant, the disputes procedure may be invoked.

Loss and expense
- Applicable procedures are covered under standard or in-house forms of contract relevant to the specific project.

Indemnities, insurances and warranties

Relevant provisions depend on the nature of the project. However, they are usually governed by the conditions specified by the forms of 'model' contracts/agreements issued by professional bodies or those in common use in the construction industry. Typical examples of insurances applicable to construction projects include:

- Contractors' all-risk policies (CAR), usually covering loss or damage to the works and the materials for incorporation into the works; the contractor's plant and equipment including temporary site accommodation; the contractor's personal property and that of employees (e.g. tools and equipment). The CAR policy is normally taken out by the contractor but should insure in the joint names of the contractor and the client (employer). Subcontractors may or may not be jointly insured under the CAR policy.

- Public liability policy – this insures the contractor against the legal liability to pay damages or compensation or other costs to anyone that suffers death, bodily injury or other loss or damage to their property by the activities of the contractor.

- Employers' liability policy – Every contractor will have this either on a company-wide basis, covering both staff and labour, or on a separate basis for the head office and for each site separately.

- Professional indemnity (PI) – The purpose of this is to cover the liabilities arising out of 'duty of care'. Typically, consultants (including the project manager) will require this policy to cover their design or similar liabilities and liabilities for negligence in undertaking supervision duties. In case of a design and build contract, the contractor has to take out a separate PI policy, as designing is not covered by the normal CAR policy.

- Integrated project insurance – Collaborative form of insurance is being increasingly used where all the key parties collectively subscribe to a single insurance policy covering all aspects of the project.

Design coordination

The project brief will be reviewed jointly by client and project manager with the aim of confirming that all relevant issues have been considered. These may include:

- health and safety obligations

- environmental requirements

- loading considerations

- space and special accommodation requirements

- eventual user's needs

- standards and schedule of finishes

- site investigation information/data

- availability of necessary surveys and reports

- planning consent and statutory approvals

- details of internal and external constraints

The project manager will need to seek the client's approval to issue the brief and relevant information to the design team and other consultants. Among the other duties which fall to the project manager are the following:

- Defining the roles and duties of the project team members.

- Responsibility for the drawings and specifications:

  - establishing format (e.g. CAD compatibility issues) sizes and distribution and seeking comments on their content and their timing

  - the issue of tender drawings and specifications

  - advising contractors and subcontractors of the implications of the design

  - setting requirements for (1) shop/fabrication drawings; (2) test data and (3) samples and mock-ups

- Monitoring the production of the outline proposals for the project by:

  - reviewing sketch plans and outline specifications in terms of the brief

  - preparing the capital budget and reconciling this with the outline budget

  - appraising the implications of the schedule

  - effecting reconciliation with the project master schedule

  - finalising the outline proposals making recommendations/presentations to the client and seeking the latter's approval to proceed

- Monitoring progress of the design work at the pre-tender stage by:

  - reviewing with the consultants concerned the client's requirements, brief documentation and their sectional implications

  - agreeing team members' input and identification of items needing client's clarification

  - reviewing with the client any discrepancies omissions and misunderstandings, seeking their resolution and confirming to the team

- Agreement of overall design schedule and related controls.

- Identification of items for pre-ordering and long delivery preparation of tender documents, client's approvals and placement of orders and their confirmation.

- Monitoring production of drawings and specifications throughout the various stages of the project and their release to parties concerned.

- Arranging presentations to the client at appropriate stages of design development and securing final approval of tender.

Change management

- Reviewing with the design team and other consultants any necessary modifications to the design schedule and information required schedules (IRS) in the light of the appointed contractors'/subcontractors' requirements and reissuing revised schedule/IRS.

- Preparing detailed and specialist designs and subcontract packages including bills of quantities.

- Making provision for adequate, safe and orderly storage of all drawings, specifications and schedules including the setting up of an effective register/records and retrieval system.

The project manager must ensure that the client is fully aware that supplementary decisions must be obtained as the design stages progress and well within the specified (latest) dates in order to avoid additional costs. Designs and specifications meeting the client's brief and requirements are appraised by the quantity surveyor for costs and are confirmed to be within the budgetary provisions.

Handling changes will require a series of actions. The project manager will be responsible for these activities:

- Administering all requests through the change order system (see Table 4.4 and Briefing Note 3.13 for checklist and specimen form).

- Retaining all relevant documentation.

- Producing a schedule of approved and pending orders which will be issued monthly.

- Ensuring that no changes are acted on unless formally decided.

- Considering amendments and alterations to the schedules and drawings within the provisions of the applicable contract/agreement.

- Initial assessment of any itemised request for change made by the client taking due account of the effect on time.

Action by consultants in relation to variations will include the following items:

- Securing required statutory/planning approvals and cost-checking revised proposals. Confirmation of action taken to project manager.

- Design process and preparation of instructions to contractors involved.

- Cost agreement procedure for omissions and additions, that is, estimates, disruptive costs, negotiations and time implications.

Site instructions

Site instructions must be issued in writing and confirmed in a similar manner by recipients. Site instructions which constitute variations can be categorised as:

- normal

- special (e.g. concerned with immediate implementation as essential for safety, health and environmental protection aspects)

- extension of time required or predicted

- additional payments involved or their estimate

Site instructions will be binding if they are issued and approved in accordance with the contract provisions.

Cost control and reporting

The quantity surveyor has overall responsibility for cost monitoring and reporting with the assistance of and input from the design team, other consultants and contractors. Action at the pre-construction stage involves the following items:

- The preparation of preliminary comparison budget estimates.

- The agreement of the control budget with the project manager.

- Project budget being prepared in elemental form; the influence of grants is identified.

- The establishment of work packages and their cost budgets.

- Costing of change orders.

Other elements associated with work control are as follows:

- Assessment of cost implications for all designs, including cost comparison of alternative design solutions.

- Value analysis procedures, including cost in use.

- Comparison of alternative forms of construction using data on their methodology and costs.

- Comparison of cost budgets and tenderers' prices at subcontract tender assessment.

- Tenders which are outside the budget and which require an input from the project manager on such matters as:

  - alteration of specifications to reduce costs

  - acceptance of tender figure and accommodating increased cost from contingency; alternatively the client may accept the increase and seek savings from other areas

  - possible retendering by alternative contractors

- Production of monthly cost reports, including:

  - variations since last report, incorporating reasons for costs increase/decrease

  - current projected total cost for the project

  - cash flow for the project: (1) forecast of expenditure; (2) actual cash flow as schedule monitoring device indicating potential overspending and any areas of delay or likely problems.

The report should be agreed with and issued to the project manager who will:

- give advice and initiate action on any problems that are identified

- arrange distribution of copies according to a predetermined list

Planning schedules and progress reporting

Planning is a key area and can have a significant effect on the outcome of a project. The handbook will set out the composition and duties of the planning support team and the appropriate techniques to be used (e.g. bar charts, networks). The planning and scheduling will then follow the steps set out in Briefing Note 3.06:

- Preparation of an outline project schedule, which will include coordination of design team contractors and client's activities then seeking of the client's acceptance.

- Production of an outline construction schedule indicating likely project duration and the basis for determining the procurement schedule.

- Production of an outline procurement schedule including the latest date for placement of orders (materials equipment contractors) and design release dates.

- Modifications if necessary to the outline construction schedule due to constraints.

- Production of the outline design schedule including necessary modifications due to external limitations.

- Preparation of the project master schedule.

- Preparation of a short-term schedule for the pre-construction stage; this will be reviewed monthly.

- Production of a detailed design schedule in consultation with and incorporating design elements from the design team members concerned including:

  - scheme design schedule

  - drawing control schedule

  - client decision schedule

  - agreement by client consultants and project manager

- Reviewing the outline procurement schedule and its translation into one, which is detailed.

- Preparation of a works package schedule.

- Production of schedules for bills of quantities procurement including identification of construction phases for tender documentation and production of tender documentation control.

- Expansion of the outline construction schedule into one, which is detailed.

- Preparation of schedules for:

  - enabling works

  - fitting out (if part of the project)

  - completion and handover

  - occupation/migration (if part of the project)

Progress monitoring and reporting procedures should be on a monthly basis and agreed following consultation with consultants and contractors. Reports will need to be supplied to the project manager who will report to the client.

Meetings

Meetings are required to maintain effective communications between the project manager, project team and the other parties concerned, for example, those responsible for industrial relations and emergencies as well as the client. The frequency and location of meetings and those taking part will be the responsibility of the project manager. Meetings held too frequently can lead to a waste of time whereas communications can suffer where meetings are infrequent. Briefing Note 3.10 contains details of typical meetings and their objectives.

Procedures for meetings include:

- agenda – issued in advance stating action/submissions required

- minutes and circulation list (time limits involved); Briefing Note 3.09 contains examples of agenda and minutes

- written confirmation and acknowledgement of instructions given at meetings (time limit involved)

- reports/materials tabled at meetings to be sent in advance to the chair

Selection and appointment of contractors

The project manager as the client's representative has the responsibility with the support of relevant consultants for the selection and appointment of:

- contractors, for example, main, management, design and build

- contractors, for example, specialist, works, trade.

The various processes associated with this activity are summarised as follows:

- Selection panel appointments relevant to the nature and scope of tender to be awarded. Nomination of a coordinator (contact) for all matters concerned with the tender.

- Establishment of selection/appointment procedures for each stage.

Pre-tender

Pre-tender activities will include the following:

- Assessment of essential criteria/expertise required for a specific tender.

- Preparation of long (provisional) list embracing known and prospective tenderers.

- Checks against database available to project manager, especially financial viability and quality of past and current work; possible use of telephone questionnaire to obtain additional data.

- Potential tenderers invited to complete/submit selection questionnaire; short list finalised accordingly.

- Arrangements for pre-qualification interview including prior issue of the following documentation relevant to the project to the prospective tenderer with interview agenda outline of special requirements and expected attendees to cover:

  - general scope of contract works and summary of conditions

  - preliminary drawings and specifications

  - summary of project master and construction schedules

  - pricing schedule

  - safety, health and environmental protection statement

  - labour relations statement

  - quality management outline

- Tender and reserve lists finalised.

Tendering process

The tendering process includes the following activities:

- Selected tenderers confirm willingness to submit bona fide tenders. Reserve list is employed in the event of any withdrawals and selection made in accordance with placement order.

- Tender documents issued and consideration given by both parties to whether mid-tender interview is required or would be beneficial.

- Interview arranged and agenda issued.

Carry out the following on the receipt of all tenders:

- evaluation of received tenders

- arrangements for post-tender interview and prior issue of agenda

- final evaluation and report

- pre-order check and approval to place order

Safety, health and environmental protection

The handbook should draw attention to the specific and onerous duties of the client and other project team members under the CDM Regulations, and should include procedures to ensure they cannot be overlooked. It is the responsibility of the principal contractor to formulate the health and safety plan for the project to be adhered to by all contractors in accordance with the CDM Regulations and taking account of other applicable legislation. Contractors are required as part of their tender submission to provide copies of their safety policy statement which outlines safe working methods that conform to the CDM Regulations.

Other matters which come within the remit of the principal contractor are as follows:

- The establishment and enforcement within the contractual provisions of rules, regulations and practices to prevent accidents, incidents or events resulting in injury or fatality to any person on the site, or damage or destruction to property, equipment and materials of the site or neighbouring owners/occupiers.

- Arranging first-aid facilities, warning signals and possible evacuation as well as the display of relevant notices posters and instructions.

- Instituting procedures for:

  - regular inspections and spot checks

  - reporting to the project manager (with copies to any consultants concerned) on any non-compliance and the corrective or preventive action taken

  - hazardous situations necessitating work stoppage and in extreme cases closedown of the site

Quality assurance: Outline

This is applicable only if quality assurance (QA) is operated as part of contractual provisions. It is critical for the client to understand the operation of a QA scheme, its application and limits of assurance and the need for defects insurance. Procedures and controls will need to be established to ascertain compliance with design and specifications and to confirm that standards of work and materials quality have been attained. The consultants will review details of their quality control with the project manager. The contractors' quality plan will indicate how the quality process is to be managed, including control arrangements for subcontractors.

Responsibility for monitoring site operation of QA administration and control procedures for the relevant documents will need to be established.

As an alternative to QA, any procedures for the management of quality should be included in the handbook (see under 'Quality management' in stage 3).

Disputes

Procedures for all parties involved in the project in the event of disagreement and disputes are to be specified in accordance with the contractual conditions/provisions which are applicable.

Signing off

Any procedures for signing-off documents should be specified. Signing-off points may occur progressively during stages of the project and be incorporated in a 'milestone schedule'. Details should include permitted signatories and a distribution list.

Reporting

The following reports are examples of what might be prepared.

Project manager's progress report

To be issued monthly and include details of:

- project status:
    - updated capital budget
    - accommodation schedule
    - authorised change orders during the month
    - other relevant matters
- operational brief
- design development status
- cost plan status and summary of financial report
- schedule and progress:
    - design
    - construction
- change/variation orders
- client decisions and information requirements
- legal and estates
- facilities management
- fitting out and occupation/(migration) planning
- risks and uncertainties
- update of anticipated final completion date
- distribution list

Design team's report

Issued monthly and including input from consultants and containing details on:

- design development status

- status of tender documents
- information produced during the month
- change orders/design progress
- information requirements/requests status
- status of contractor/subcontractor drawings/submittals
- quality control
- distribution list

Financial control report

Issued monthly and including:

- reconciliation capital sanction/capital budget
- updated cost plan and anticipated final cost projection
- authorised change orders – effects
- pending change orders – implications
- contingency sum
- cash flow
- VAT
- distribution list

Daily/weekly diary

Prepared by each senior member of the project team and filed in its own separate loose-leaf binder for quick reference and convenient follow-up. Diaries are made accessible to the project manager and typically contain:

- a summary of forward and ad hoc meetings and people attending
- a summary of critical telephone conversations/messages
- documents received or issued
- problems comments or special situations and their resolution
- schedule status (e.g. work package progress or delays)
- critical events and work observations
- critical instructions given or required
- requests for decisions or actions to be taken
- an approximate time of day for each entry
- a distribution list

Construction stage

The handbook will include procedures for the following activities:

- Issuing drawings, specifications and relevant certificates to contractors.
- Actioning the consultants' instructions, lists, schedules and valuations.

- Aspects prior to commencement such as:

  - recording existing site conditions, including adjacent properties

  - ensuring that all relevant contracts are in place and that all applicable conditions have been met

  - confirming that all risk insurance for site and adjacent properties is in force

  - ensuring that all site facilities are to the required standard including provisions for health, safety and environmental protection

- Control of construction work including:

  - reviewing a contractor's preliminary schedule against the master schedule and agreeing adjustments

  - ensuring checks by the main contractor on subcontractor schedules

  - checking and monitoring for all contractors the adequacy of their planned and actual resources to achieve the schedule

  - approvals for subletting in accordance with contractual provisions

  - reporting on and adjusting schedules as appropriate

  - checks for early identification of actual or potential problems (seeking client's agreement to solutions of significant problems)

- Controls for variations and changes (see Briefing Note 3.13).

- Controls for the preparation and issue of change orders (see Figure 4.4 and Briefing Note 3.13).

- Processing the following applications for the client's action:

  - interim payments from consultants and contractors

  - final accounts from consultants

  - final accounts from contractors subject to receipt of relevant certification

  - payment of other invoices

- Making contact and keeping informed the various authorities concerned to facilitate final approvals.

- The design team and other relevant consultants to supervise and inspect works in accordance with contractual provisions/conditions and participate in and contribute to:

  - the monitoring and adjustment of the master schedule

  - controls for variations and claims

  - identification and solutions of actual or potential problems

  - subletting approvals

  - preparation of change orders

Operating and maintenance

The procedures for fitting out should be designed to avoid divided responsibility in the case of failure of parts of the building or its services systems. The procedures to be used in the handbook can be developed by reference to the relevant sections in stages 5–7. They should include adequate arrangements for the management of any interfaces between contracts or work packages. It is especially vital to have procedures for:

- the transition of commissioning data record drawings and operating and maintenance manuals from one contract to the next

- confirmation that all relevant handover documentation and certification has been completed

Testing and commissioning

Testing and commissioning is part of the construction stage. It is the main contractor's responsibility which is delegated to the services subcontractors. Action is taken in two stages: pre-contract and contract/post-contract.

Pre-contract

Ensuring that the client recognises commissioning as a distinct phase of the construction process starting at the strategy stage:

- Ensuring that the consultants identify all services to be commissioned and defining the responsibility split for commissioning between designer, contractor, manufacturer and client.

- Identifying statutory and insurance approvals required and planning to meet requirements and obtain approvals.

- Coordinating the consultants' and client's involvement in commissioning to ensure conformity with the contract arrangements.

- Arranging single-point responsibility for control and the client's role in the commissioning of services.

- Ensuring contract documents make provision for services commissioning.

Contract and post-contract

Ensuring relevant integration within construction schedules:

- Monitoring and reporting progress and arranging corrective action.

- Ensuring provision and proper maintenance of records, test results, certificates, checklists, software and drawings.

- Arranging for or advising on maintenance staff training, post-contract operation and specialist servicing contracts.

Examples of a checklist and documentation are given in Briefing Notes 7.03 and 7.04.

Completion and handover

The closely interlinked processes of completion and handover are very much a hands-on operation for the project manager and his team. This stage provides the widest and closest involvement with the client. Completion and handover require careful attention because they determine whether or not the client views the whole job as successful.

Completion

Handbook procedures may cover two types of agreement:

- Agreements for partial possession and phased (sectional) completion (if required):

  - access inspections, defects, continuation of other works and/or operation of any plant/services installation material, obstructions or restrictions

  - certification on possession of each phase; responsibility for insurance

- Agreements and procedures associated with practical completion:

  - user/tenant responsibility for whole of the insurance

  - provision within a specified time limit of complete sets of as-built and installed drawings, mechanical and engineering and other relevant installations/services data as well as all operating manuals and commissioning reports

  - storage of equipment/materials except those required for making good any defects

  - access for completion of minor construction works, rectification of defects, testing of services, verification of users' works and other welfare and general facilities

Briefing Note 7.06 provides a typical checklist at the practical completion stage.

Handover

Procedures are needed for the following activities:

- To ensure that handover only takes place when all statutory inspections and approvals have been satisfactorily completed and subsequently to arrange that all outstanding works and defects are resolved before expiry of the defects liability period.

- To provide and agree a countdown schedule with the project team (examples of handover inspections and certificates checklists are given in Briefing Note 7.05).

- To define responsibilities for all inspections and certificates.

- To monitor and control handover countdown against the schedule.

- To control pre-handover arrangements if the client has access to the building before handover.

- To implement the pre-agreed Soft Landing arrangement for a smooth handover and transition (see Briefing Note 1.05)

- To deal with contractors who fail to execute outstanding works or correct defects including the possibility of implementing any contra-charging measures available under the contract. Agree and set up a procedure for contra-charging.

- To monitor and control any post-handover works which do not form part of the main contract.

- To monitor and control outstanding post-completion work and resolution of defects which form part of the main contract.

- To manage the end of the defects liability period and implement relevant procedures.

- To establish arrangements for the final account issuing the final certificate and carrying out the post-completion review/project evaluation report.

- To facilitate a Post occupation evaluation has been undertaken to determine the performance and success criteria of the project.

- To facilitate transfer of information and ownership of BIM from the delivery team to the operation and maintenance team

Client commissioning and occupation

Client commissioning

Client commissioning will involve the following handbook procedures:

- Arranging the appointment of the commissioning team in liaison with the client and establishing objectives (time, cost and specifications) and responsibilities at the feasibility and strategy stages.

- Preparation of a comprehensive commissioning and equipment schedule.

- Arranging access to the works for the commissioning team and client personnel during construction including observation of engineering services commissioning.

- Ensuring coordination and liaison with the construction processes and consultants.

- Preparing new work practice manuals and in close liaison with the client's/user's facilities management team, arranging staff training and recruitment/secondment of additional staff (e.g. aftercare engineer to support the client during the initial period of occupancy).

- Deciding the format of commissioning test and calibration records.

- Renting equipment to meet short-term demands.

- Deciding quality standards.

- Monitoring and controlling commissioning progress and reporting to the client.

- Reviewing post-contract the operation of the building at 6, 9 and 12 months: improvements, defects, corrections and related feedback.

Briefing Note 7.01 contains a relevant checklist.

Occupation

Occupation can be part of the overall project or a separate project on its own. A decision to this effect is made at the strategy stage with the client or user. The separate stages of occupation are set out below. Figure 7.1, Figure 7.2, Figure 7.3 and Figure 7.4 illustrate these procedures graphically and Briefing Note 7.01 provides an example of an occupation implementation plan. Briefing Note 8.01 provides an outline of post-occupation evaluation process which should also include a Post Completion Review to assess the performance and success of the project.

Structure for implementation   In order to achieve the necessary direction and consultations, individuals and groups are appointed, for example:

- project executive (client/occupier/tenant)

- occupation coordinator (project manager)

- occupation steering group of a chair, coordinator and functional representatives concerned with the overall direction for:

  - construction schedule

  - technology

- space planning

- facilities for removal

- user representation

- costs and budget outline

- senior representative meeting of a chair (functional representative on steering group), coordinator and senior representatives of a majority of employees concerned with consultations on:

  - space planning

  - corporate communications

  - construction schedule problems

  - technology

- local representative groups chaired by manager/supervisor of their own group concerned with consultation at locations and/or departmental levels in order to ensure procedures for regular communications

Scope and objectives (regularly reviewed)

- Identification of who is to move (project executive).

- Agreement on placement of people in new locations (steering group).

- Decision on organisation of move (steering group):

  - all at once

  - several moves

  - gradual flow

- Reviewing time constraints (steering group):

  - construction

  - commercial

  - holidays

- Identification of risk areas, for example:

  - construction delays and move flexibility

  - organisational changes

  - access problems

  - information technology requirements

  - furniture deliveries and refurbishment

  - retrofit requirements

Methodology

- Listing special activities needed to complete the move, for example:

  - additional building work

  - communications during move

  - provision of necessary services and move support

- corporate communications

- removal administration

- furniture procurement

- removal responsibility in each location/department

- financial controls

- access planning

- Preparation of a task list for each special activity, confirmation of the person responsible and setting the schedule of project meetings.

- Production of outline and subsequently detailed schedule.

Organisation and control

- Steering group establishes 'move group' to oversee the physical move.

- Production of 'countdown' schedule (move group).

- Identification of external resources needed (move group), for example:

  - special management skills

  - one-off support tasks

  - duplication of functions during move

- Reporting to the client external support needs and costs (steering group).

- Preparation of monitoring and regular review of actual budget (steering group), for example:

  - dual occupancy

  - special facilities

  - additional engineering and technology needs

  - planning and coordinating process

  - inflation

  - external resources

  - non-recoverable VAT

  - contingencies

Briefing Note 1.05   Government Soft Landings

Summary

The UK government has identified[3] a need to improve value in government construction contracts.

This 'Soft Landing' approach was conceived as a way forward to improve performance of facilities for their users.

The Government Soft Landings (GSL) policy was therefore produced.

Aim

To align the interests of those who design and construct a facility or asset, with those who eventually use it.

Asset management

The aim is to have an introduction period of between three and five years, post-completion (CIOB Stage 8), to confirm satisfactory in-use status of the facility, for its whole life.

To this effect, there is a strong link with BIM.

Development of the GSL

The GSL policy was developed with the assistance of a task group formed from construction and FM suppliers, architects/designers, academics and representatives from industry, local and central government.

It is aligned to the BSRIA Soft Landings Framework (BSRIA BG4/2009) and the BSRIA Soft Landings Core Principles (BSRIA BG 38/2012), later the Cabinet Office produced eight sections for the Government Soft Landings (April 2013 http://www.bimtaskgroup.org/reprts).

After GSL initially identified four sections/areas along the project timeline, these eventually have become eight sections, including Planning for aftercare:

1. Introduction

2. GSL Lead and GSL Champions

3. Functionality and effectiveness:

---

[3]   The UK Government Soft landings Policy, September 2012 – April 2013, issued by the Cabinet Office – Government Property Unit.

Buildings designed to meet the needs of the occupiers; effective, productive working environments.

4.   Capital cost and operating cost

5.   Environmental management:

Meet government performance targets in energy efficiency, water usage and waste production.

6.   Facilities management:

A clear, cost-effective strategy for managing the operations of the facility.

7.   Commissioning, training and handover

Projects delivered, handed over and supported to meet the needs of the end user

8.   Planning for Aftercare

The project manager has the responsibility of organising and reviewing the project specific Aftercare Plan, established and approved prior to commencement of details design and construction and further approved by the project sponsor prior to implementation.

Collaborative working

Collaborative working is fundamental to the concept of GSL and it is paramount to obtain key stakeholder engagement at all stages of the process.

1.05 Briefing Note

2 Feasibility

Stage checklist

Key processes: Project brief
Project manager selection
Feasibility studies
Business case
Funding options
Delivery parameters

Key objective: 'Is the *need* feasible?'

Key deliverables: Project brief
Signing off business case

Key resources: Client team
Project manager
Specialist consultants

Stage process and outcomes

During this stage, the client is establishing objectives and reviewing the various options available to achieve these objectives. Commencing with only a conceptual understanding of what the client requires this stage involves developing and defining the client's requirements, identifying and appraising all the options, and concludes with a client decision to proceed on a firm course of action.

Outcomes:

- definition of client's objectives

- appraisal of different options

- a recommended best course of action to achieve the objectives

- demonstration of the financial viability of the preferred solution

- confirmation of a project brief

- business case (produced by the client)

- client executive-level approval to project proposals (usually based on a business case)

- client decision to proceed or not to proceed to the strategy stage

- prepare core consultant scope of services

Code of Practice for Project Management for Construction and Development, Fifth Edition. Chartered Institute of Building.
© Chartered Institute of Building 2014. Published 2014 by John Wiley & Sons, Ltd.

Client's objectives

The main objectives for the client at this stage include identifying and specifying the project objectives, outlining possible options and select the most suitable option through sustainability, value and risk assessment. At this stage, establishing the project execution plan (PEP) for the selected option should be the key output.

Outline project brief

For most clients, a building is not an end in itself, but merely the means to an end: the client's corporate objectives. The client's objectives may be as complex as the introduction and accommodation of some new technology into a manufacturing facility or the creation of a new corporate headquarters; or they may be as simple as obtaining the optimum return on resources available for investment in a speculative office building.

The client's objectives are usually formulated by the organisation's board or policy-making body (the investment decision maker) and may include certain constraints – usually related to time, cost, performance and location. The client's objectives must cover the function and quality of the building or other facility.

If it is considered that the objectives are complex enough to merit the engagement of a project manager, the appointment should ideally be made as early as possible, preferably after approving the project requirements at the inception stage. This will ensure the benefit of the special expertise of the project manager in helping to define the parameters and in devising and assessing options for the achievement of the objectives.

The project manager should be provided with, or assist in, preparing a clear statement of the client's objectives and any known constraints. This is the initial outline project brief to which the project manager will then work. A typical example of a template for an outline project brief is shown in Figure 2.1.

Feasibility studies

There is seldom, if ever, a single route available for the achievement of the client's objectives, so the project manager's task is to work under the client's direction to help establish a route which will best meet the client's objectives within the perceived constraints that are set. In liaison with the client, the project manager will discuss the available options and initiate feasibility studies to determine the strategy to be adopted. In order that the feasibility studies are effective, the information used should be as full and accurate as possible. Much of that information will need to be provided by specialists and experts. Some of these experts may be available within the client's own organisation or be regularly retained by the client: lawyers, financial advisors, insurance consultants and the like. Others, such as architects, engineers, quantity surveyors, project planners, planning supervisors, town planning consultants, land surveyors and geotechnical engineers, may need to be specially commissioned. In some instances, it is desirable to involve constructors (e.g. in case of framework contracts or even design and build contracts) in preparation and completion of the feasibility study.

Feasibility study reports should include the following:

- scope of investigation (from outline project brief) including establishing service objectives and financial objectives

- studies on requirements and risks

PROJECT TITLE

PROJECT REF

CLIENT:

PROJECT SPONSOR:

PROJECT MANAGER:

GOAL

This needs to be specific and include the justification for the project

It should spell out : **what** will be done and by **when**;

OBJECTIVES

It is essential these cover the OUTCOMES expected of the project and that preferably they are:

Specific – i.e. clear and relevant

Measurable – i.e., so it is feasible to see how it is progressing

Achievable / agreed to – helpful to use positive language and that others 'buy-in' to the objectives

Realistic – this depends of three factors: resources/ time/ outcome or aim

Time bound have a time limit – without this they are wishes

APPROACH

The project plan should include the key milestones for the review, i.e., set a target date for agreeing the project brief and target dates for achieving key stages of the project.

SCOPE

This sets the project boundaries and it can be useful to add what is NOT covered.

It is an essential reference point if the project changes in due course

CONSTRAINTS

Could state 'start' date and 'end' dates, functional restraints, operational parameters, etc.

DEPENDENCIES

This identifies factors outside the control of the project manager, and may include

■ Supply of information

■ Decisions being taken at the right time

■ Other associated projects

RESOURCE REQUIREMENTS

Include estimate of input required from the project team

AGREED

Signature:

Date:

Project Manager:

Project Sponsor :

Note: The above example of an outline project brief is for guidance purposes only.

Figure 2.1 Outline project brief.

- public consultation (if applicable)

- consultation with stakeholders and third parties

- a geotechnical study (if applicable)

- environmental performance targets (e.g. BREEAM or LEED, Code for Sustainable Homes and EcoHomes; refer to Briefing Note 2.01)

- an environmental impact assessment (refer to Briefing Note 2.02)

- a health and safety study

- legal/statutory/planning requirements or constraints

- risk management strategy

- estimates of capitals and operating costs (demolition costs, if applicable)

- assessment of potential funding

- potential site assessments and remediation strategy (if applicable)

- master development schedule

The client will commission feasibility studies and establish that the project is both financially viable and deliverable. The client may instruct the project manager at this stage and, if so, his input will be made alongside the reports and views of the various consultants.

The client may ask the project manager to engage and brief the various specialists for the feasibility studies, coordinate the information, assess the various options and report on his conclusions and recommendations. The feasibility report should include as a minimum a 'risk assessment' for each option and will usually also determine the contractual procurement route to be adopted and an outline development schedule applicable to each. The client may also require comparative 'lifecycle costings' to be included for each option. The issue of sustainability is now a significant part of all development projects, whether new build or refurbishment, and the three elements of environment, economy and society must be considered from the earliest stages of each project. Further guidance to prepare the feasibility report as part of the business case is available in Briefing Note 2.05.

During the progress of the feasibility studies, the project manager will convene and minute meetings of the feasibility team, report progress to the client and advise the client if the agreed budget is likely to be exceeded. Feasibility studies including revenue assessment are the most crucial, but also the least certain phase of a project. Time and money spent at this stage will be repaid in the overall success of the project. The specialists engaged for the feasibility studies are most commonly reimbursed on a time–charge basis and without commitment to engage the specialist beyond the completion of the feasibility study, although often some or all members of the feasibility team will be invited to participate in the selection process to become design team members.

The project manager will obtain a decision from the client on which option to adopt for the project and this option is designated the outline project brief. The process of developing the project brief from the client's objectives is shown in Figure 2.2.

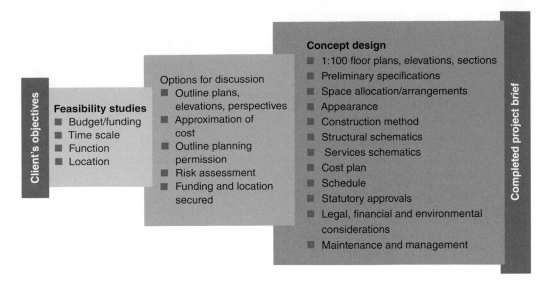

Figure 2.2 Development of project brief from objectives.

Energy in a building environment

Research[1] suggests that globally existing buildings account for approximately 40% of the world's total primary energy consumption and 24% of the world's CO_2 emissions.

Approximately 42% of all carbon emissions in the UK come from buildings.[2] The UK Government has recently introduced several measures to improve the energy efficiency of buildings, including:

- introducing energy performance certificates (EPCs) for properties providing A-G efficiency ratings and recommendations for improvement

- requiring public buildings to display energy certificates (DECs)

- requiring inspections for air conditioning systems

- giving advice and guidance for boiler users

Since October 2008 all properties – homes, commercial and public buildings – when, sold, built or rented, have needed an EPC and larger public buildings need to display a display energy certificate (DEC).

The Energy Performance of Buildings Directive stipulates the minimum requirements in terms of energy performance of buildings which all member states must adopt.

Generally, building integrated energy systems, providing heat and power, for example, photovoltaic building façades, ground source heating (or cooling) systems, integration of heating and ventilation systems, recirculating heat gain systems, and more and more utilisation of natural daylight (at times also using solar energy to generate power) are being investigated and encouraged by both clients and designers in addition to the more traditional enhancement of insulation properties to achieve compliance and often surpass the minimum requirements for energy management.

1   http://www-csd.eng.cam.ac.uk/themes0/energy-demand/energy (accessed March 2014).
2   http://www.communities.gov.uk/planningandbuilding/sustainability/energyperformance/ (accessed March 2014).

Lifecycle costing and sustainability

Lifecycle cost is defined as 'the cost of an asset, or its parts throughout its life-cycle, while fulfilling the performance requirements[3]'.

Lifecycle costing is the 'methodology for systematic economic evaluation of life-cycle costs over a period of analysis as defined in the agreed scope[4]'.

The appraisal of lifecycle costing begins at the feasibility stage, where it forms the part of the outline business case during options appraisal where it ensures that all options are being compared on a like for like basis. At the approval of the business case, the lifecycle costing should include cash-flow forecast prediction, which will have an impact on investment/funding choices. It will also help to assess the measures of economic performance, indicating financial criteria such as payback and internal rate of return.

The key but simple questions that need to be considered for lifecycle costing are

- What is needed now and how much will it cost?
- How much will be needed in the future to maintain the 'need'?
- How long is the future?
- What is the best way to evaluate future costs against current costs?

The concepts central to the green/sustainability movement harmonises into lifecycle costing. Procuring authorities that strive to create and operate high-performance, environmentally friendly, energy-efficient, healthful facilities are finding that their choices are resulting not only in exceptional environments, but also long-lasting facilities that will use fewer resources and result in measurable long-term savings.

Lifecycle costing ensures investors and decision makers to evaluate the entire cost of a facility, not just the initial delivery cost. Because costs associated with a facility do not stop when construction is complete, the costs to operate, maintain and dispose of a facility over its useful life must be considered. By making informed decisions and investing a little more at the early stages during planning, procuring authorities can realise significant savings over the life of a building.

At the completion of the facility, during post-occupancy evaluation (see Briefing Note 8.01) information collected will assist to benchmark the achievements of the key decisions taken during the lifecycle costing assessment.

Sustainability in the built environment

The impact on the environment of construction and related activities is now well publicised and governments around the world are challenging construction processes, methods, practices and procedures and are monitoring the industry's outputs. The aim is to both question and change the way modern buildings are constructed and used. It is expected that such changes to processes and so forth will have a great impact on both embodied energy and also on energy input requirements of buildings, both existing and new and thus positively impact global emissions of greenhouse gases. Such change will not come cheap as construction processes are adapted and building standards and regulations set new challenges to meet future human requirements and governmental goals. In addition, through *Development Control*,* there are strict legislative requirements to consider other sustainable aspects including the balance of nature (e.g. water displacement and usage, arboriculture issues, ecological issues and similar) for any development work that require a planning application.

---

[3]   ISO 15686–5:2008 Buildings and constructed assets – Service-life planning – Part 5 life-cycle costing.
[4]   ISO 15865–5:2008 Buildings and constructed assets – Service-life planning – Part 5 life-cycle costing.
*   See glossary for definition.

2 Feasibility

Towards sustainable development

There is a tendency by many people to consider 'sustainability' as just the 'environment', perhaps because there is so much written about climate change and the emphasis on reducing carbon emissions and pollution. It is these drivers that appear to be leading the sustainability agenda. However, the other two important elements of the 'triple bottom line' which comprise 'sustainability' are social (people) and economic (profit). Of these it is the latter – economic – which many believe is the greater concern, for saving the environment is not expected to be cheap and it is only natural that most stakeholders want to protect their profits. The National Planning Policy Framework (NPPF) (2012)[5] is placing increasing importance on the impact of environmental changes.

With an increased world population and diminishing resources, even in the absence of climate change, there remains a need to use materials more wisely and reduce the amount that goes to landfill. The reuse or recycling of valuable resources therefore makes environmental as well as economic sense.

The Stern report (2006)[6] believes extreme weather, created by climate change, could reduce global GDP by up to 1% with the worse-case scenario predicting global output per head falling by 20%. This cost of 'doing nothing', however, can arguably be much greater than the costs associated with reducing carbon emissions, although there are many who show reluctance to the implementation of the preventative actions needed to reduce such emissions, on the grounds of associated costs and hence loss of profit.

However, many consultants are successfully demonstrating that if sustainability is approached holistically, using integrated design with modern materials and systems, it is entirely possible to create a built environment that is both sustainable and economically viable. Furthermore, the argument now held by the majority of industry experts is that we really have no choice, the planet is in need of environmental adjustment and the only way for this to happen is by reducing carbon emissions through changes to our built environment, our energy sources and consumptions, transport systems and lifestyles. But such changes are not simple in today's world of politics and limited resources, consequently there is a need to consider holistically the 'triple bottom line' of environment, economic and social, which means that profit and people must remain as key considerations of the overall solution.

Design should be such that should function change the facility can be adapted with relative ease utilising the principle of long life loose fit, thus reducing the use of material resources.

In terms of social (people) acceptance, the construction industry needs to build facilities which do not adversely affect the external environment through pollution, excessive size, unsightly design, overcrowding, high maintenance and indulgent consumption of materials and resources. At the same time, homes must be affordable and available, with adequate communal green areas and spaces within environments which are comfortable in terms of light, noise, density and within reasonable commuting distances of workplaces. Travel is a key source of carbon pollution and natural resource consumption, hence creating shorter commuting distances is part of the overall aims for developing a healthier environment.

Offices and shops should follow similar guidelines by being socially acceptable and within a built environment which encompasses safety, ease of access to transport and entertainment facilities, with well-designed structures and community engagement facilities designed to create healthy lifestyles and spiritual well-being.

---

[5] Department of Communities and Local Government (2012) National Planning Policy Framework – available at https://www.gov.uk/government/uploads/system/uploads/attachment_data/file/6077/2116950.pdf

[6] Stern, N. (2006). "Stern Review on The Economics of Climate Change". HM Treasury, London. Available at http://webarchive.nationalarchives.gov.uk/+/http://www.hm-treasury.gov.uk/independent_reviews/stern_review_economics_climate_change/stern_review_report.cfm

2 Feasibility

With regard to the economic (profit) element, making buildings and facilities financial viable while being environmentally compatible is the ultimate goal for the construction industry. Unless there is careful coordination of the design, manufacture, choice of materials and specification, construction and operational efficiency, it can be very difficult to create environmentally and socially acceptable buildings within realistic budget limits. This is particularly true when refurbishments are undertaken.

If we wish to maintain the current balance of nature, we must abandon wasteful practices and produce facilities which are sustainable, socially acceptable and which protect our natural resources without destroying the environment. However, it is important to create comfortable internal environments which are constructed and operated efficiently, while producing financial returns which are also sustainable. To achieve such harmonious environments it is important to follow the national and internationally agreed standards, that is, BREEAM, LEED, GreenStar, etc. This will help reduce consumer demand for heavily polluting goods and services, the aim being to promote cleaner energy and transport systems with non-fossil fuels producing at least 60% of the required energy output by 2050 in order to achieve the required drastic reductions of carbon emissions. It could also be suggested that user/client/organisational behaviour/expectations need to change significantly to achieve this.

Responsible sustainable development

It is anticipated that eventually governments are likely to insist on carbon footprint measurements for each project period, bearing in mind that major projects can take many years to complete. The selection of materials such as concrete and steel, which have high carbon manufacturing consumption as well as potentially high transportation costs, might well bring about the use of alternative materials. Systems such as mechanical and electrical equipment and cladding may well be carbon assessed and their location of manufacture, their indirect and embodied carbon including transportation and the operational carbon consumptions, are likely to be similarly assessed and weighted against initial capital costs, when determining the choice of supplier.

Responsible sustainable development has therefore become a growing national and international issue and project managers and clients should cooperate to make sure that the whole supply chain is aware of its duties during the early stages of each project. The project manager, in particular, needs to assure the client that proper consideration by the design and construction teams will be given to the sustainability aspects of the project and its construction process so that the final building will have a minimal detrimental effect on the natural environment. To help achieve this, the project team will need to adopt innovative methods and best practice. Therefore, controlling the construction process by effective use of advanced management tools and systems (see later text) should consider the carbon footprint of the whole construction process from inception to completion, although efficient monitoring of such tools will need to be undertaken along with the measurement of the associated impact being experienced.

Typical management tools and systems include the following:

* Consider the functional life of materials and whether they can be reused or recycled and also their contribution to thermal performance of the facility during the facility lifecycle. Materials with an initial low embodied energy, during the life of the facility, if need to be replaced frequently then over the lifecycle of the building may be environmentally more damaging.

- Consider material sourcing and how locally materials are sourced.
- Consider if there is a supply chain partnering that allows return of surplus materials and/or packaging.
- Consider use of reused and recycled materials, and end use of all components.
- Detailed site waste management plans (SWMPs) with emphasis on minimum waste generation on all projects, in particular those over £300,000.
- Use of competent ICT with a top-down approach to managing sub-contractors via the use of inter-operable software systems.
- Use of electronic document management systems with secure databases.
- Incorporation of building information modelling (BIM) system approaches to design from an early stage in the process.
- The use of e-tendering and e-commerce solutions approaches to project management in the pre-construction and construction phases.
- Installation of effective wireless technologies and Radio-frequency identification (RFID) data collection devices.
- Instant messaging and the use of virtual office(s) and conferencing facilities as a means of reducing the need for travel between project team members.
- Ensuring accurate whole-life costs (WLC) are applied to the project building so that clients can make the best judgements at all stages.

Achieving sustainable development

Sustainable development in terms of design and construction imposes duties on all those involved on the entire project at all stages from inception to completion and then through the life of the development to the grave (and then back again).

Briefing Note 2.01 provides further information on key sustainability issues and actions at each project stage and Briefing Note 2.02 provides further information on BREEAM, Code for Sustainable Homes and EcoHomes.

A major consultation and review process is being undertaken by the Government, housing standards review, the outcome of which is likely to have impact on the building regulation framework of voluntary housing standards.

Most clients and government departments are proactively engaged and committed to securing sustainable development. They require the construction industry to respond to the greater demand for social, economic and environmental improvements such as living within the planet's environmental resources and creating biodiversity, ensuring a strong, healthy and just society with equal opportunities for all, building a strong and stable economy, using sound science for advancement and promoting good governance.

Sustainable development (or sustainability) is defined in BS8900 as 'an enduring, balanced approach to economic activity, environmental responsibility and social progress'. In CIBSE's introduction to sustainability, it is about enabling: 'all people throughout the world to satisfy their basic needs and enjoy a better quality of life without compromising the quality of life for future generations'. In the Report of the World Commission on Environment and Development, *Our Common Future* (The Brundtland Report) (1987),[7] it is the development 'that meets the needs of the present without compromising the ability of future generations to meet their own needs'.

Sustainable development is frequently defined as the interaction of social, economic and environmental (ecological) issues. This is represented in Figure 2.3.

[7] World Commission on Environment and Development (1987). Our Common Future (also known as the Brundtland Report). Oxford: Oxford University Press. ISBN 019282080X. Available at http://www.un-documents.net/our-common-future.pdf

2 Feasibility

Figure 2.3 A summary of sustainable development. Adapted from CIRIA C571 'Sustainable construction procurement: a guide to delivering environmentally responsible projects'.

Site selection and acquisition

Site selection and acquisition is an important stage in the project cycle where the client does not own the site to be developed. It should be effected as early as possible and, ideally, in parallel with the feasibility study. (It is to be noted that the credibility of the feasibility study will depend on the major site characteristics.) The work is carried out by a specialist consultant and lawyers and may involve a substantial due diligence exercise. This will be monitored by the project manager.

The objectives are to ensure that the requirements for the site are defined in terms of the facility to be constructed, that the selected site meets these requirements and that it is acquired within the constraints of the outline development schedule and with minimal risk to the client.

To achieve these objectives the following tasks will need to be carried out:

- Preparing a statement of objectives/requirements for the site and facility/buildings and agreeing this with the client

- Preparing a specification for site selection and criteria for evaluating sites based on the objectives/requirements

- Establishing the outline funding arrangements

- Determining responsibilities within the project team (client/project manager/ commercial estate agent)

- Appointing/briefing members of the team and developing a schedule for site selection and acquisition; monitoring and controlling progress against it

- Actioning site searches and collecting data on sites, including local planning requirements, for evaluation against established criteria

- Evaluating sites against criteria and producing a shortlist of three or four; agreeing weightings with the client

- Establishing initial outline designs and developing costs

- Discussing short-listed sites with relevant planning authorities

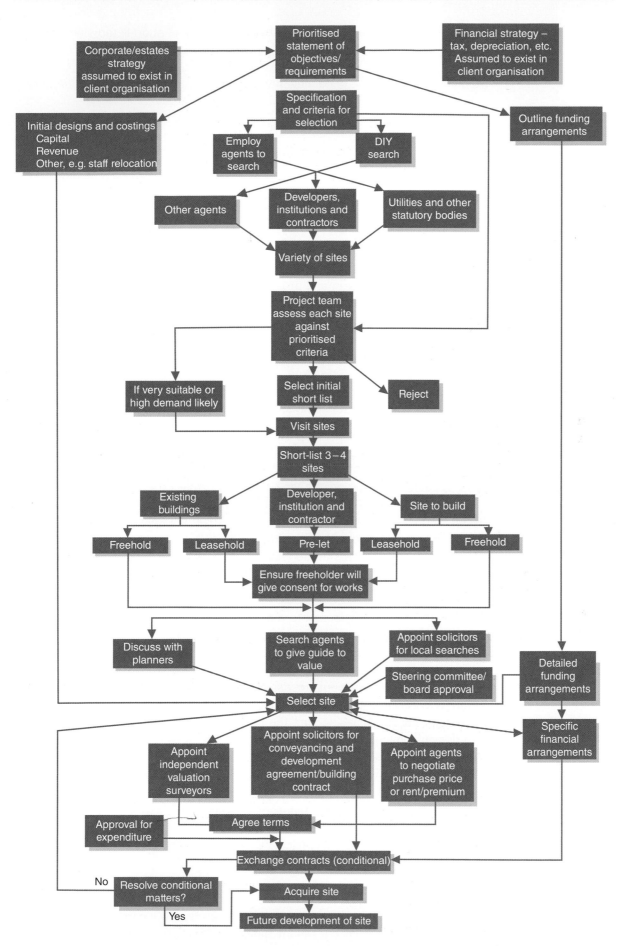

Figure 2.4 Site selection and acquisition.

- Obtaining advice on approximate open-market value of short-listed sites

- Selecting the site from a shortlist

- Appointing agents for price negotiation and separate agents for independent valuation

- Appointing solicitors as appropriate

- Determining specific financial arrangements

- Exchanging contracts for site acquisition once terms are agreed, conditional on relevant matters, for example, ground investigation, planning consent (Figure 2.4)

Project brief

The formulation of the brief for the project is an interactive process involving most members of the design team and appropriate representatives of the client organisation. It is for the project manager to manage the process, resolving conflicts, obtaining client's decisions, recording the brief and obtaining the client's approval. Managing the client organisation to ensure input into the brief comes from the right person is as important as it is time consuming. This ensures that the project manager gets to know and understand the structure, culture and personalities of the client body. At best this establishes ownership and champions for various aspects of the project at an early stage. Table 2.1 lists some suggested contents for a project brief.

If earlier work has been done, the project brief may refer to the document(s) containing useful information, such as the outline project brief, rather than include copies of them. It is not unusual during this phase for the client to modify his thinking on various aspects of the proposals, and there is certainly the opportunity and scope for change during this phase. Figure 2.5 demonstrates graphically the relationship between 'scope for change' and the 'cost of change' set against the time-scale of a development. It will be seen that the crossover point occurs at the completion of the strategy stage. The client's attention should always be drawn to this relationship and to the benefits of brief and design freezes.

The key emphasis for the client should be to understand and establish enough information about the end requirements and objectives for developing the project. This point cannot be overemphasised. It is essential that the project manager identifies the client's needs and objectives through careful and tactful examination, in order to minimise the risk of potential future changes to the project brief. Many clients who are unfamiliar with the development process are not perhaps fully aware of the importance of getting the design right as much as feasible at the start of the project, and therefore the importance of making sure the brief fully reflects the client's requirements, before design commences and getting the design right before materials ordering and construction commences. The project manager should do his utmost, therefore, to familiarise the client with the potential cost and time implications of design changes and identify as clearly as possible the precise requirements of the client. While this input may come from the project manager, it is advisable, especially for complex and business critical projects, that a 'client advisor' is appointed to give independent advice until the need for a construction project has been established.

Design brief

Within the project brief, the assembly of the design brief will normally be the responsibility of the lead design consultant along with the project manager and, where appropriate, the client and the constructors. The project manager will monitor the

Table 2.1 Contents for project brief

The following is a suggested list of contents, which should be tailored to the requirements and environment of each project.

- **Background**
- **Project definition, explaining what the project needs to achieve. It will contain**
  1. Project objectives
  2. Project scope
  3. Outline project deliverables and/or desired outcomes
  4. Any exclusions
  5. Constraints
  6. Interfaces
- **Outline business case**
  1. A description of how this project supports business strategy, plans or programmes
  2. The reason for selection of this solution
- **Funding objectives**
- **Performance expectations**
- **Acceptance criteria**
- **Risk assessment**

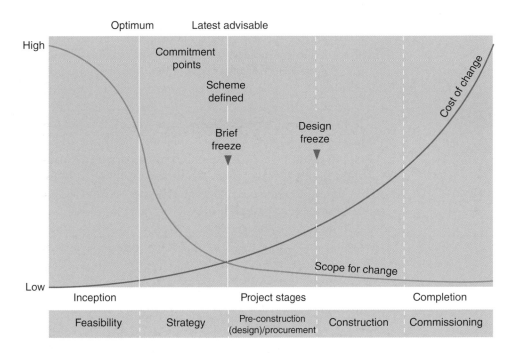

Figure 2.5 Relationship between scope for change and cost of change.

assembly of the design brief to ensure compliance with the outline project brief, the project budget and the master programme.

Depending on the nature of the project, procurement method adopted and the master programme, it is sometimes practical for some activities to proceed in parallel. However, if the design brief is not finalised before the final concept design is started, then change, delay and increased costs are almost bound to follow. The site preparation should never start until after the concept design has been finalised and signed off. While decisions on some elements of the design brief may be deferred by the client even until after construction has commenced, this inevitably involves some risk for which time and cost contingencies should be provided. Accordingly, in order to promote certainty, it is much better and usually possible (except in extreme circumstances) to complete the brief and the design before construction begins. It just needs good project management.

The project manager will advise the client of the implications for cost, time and risk in the deferment of any elements of the design brief. The project manager will monitor the progress of the assembly of the concept design and notify the client of the effects on cost, time, quality, function and financial viability of any changes from the design brief.

Funding and investment appraisal

In all development projects, a balance between cost and value must be established. The financial appraisal of the project can either be assessed by calculating the total cost and then assessing the value or alternatively, calculating the value of the end product and working out the project costs with an eye to value. In either case, the client will expect value to exceed cost, and in the case of developer-led projects the client will at inception stage have decided on the level of profit (or benefit) required for the amount of risk involved. A thorough risk analysis, particularly analysing the market conditions on the potential revenue generation, interest rate changes, potential impact of programme delay and outcomes of similar historical precedents (which may be incorporated within the business plan or development appraisal for the project), is usually performed to assist in decision making. Developers and many clients experienced in construction procurement may not require specific help from the project manager in these areas but should keep him well informed of the financial arrangements so that they can be taken into account in any project decisions. On the other hand, clients unfamiliar with construction may require an input by the project manager or from another independent advisor. In any case, although the project manager may have knowledge of project finance, it is unlikely that he will be expected to advise in this area. Specialist advisors or the client himself will arrange bank finance; take tax and legal advice in all those areas relating to the acquisition of the site and the financing of the development project. The project manager should be able to advise on certain matters relating to VAT, budgetary systems, cost and cash flow. The project manager should also know when and where to go for specialist advice to augment his own expertise or his client's expertise in such matters.

Development planning and control

Planning has become a key means of delivering a number of the government's objectives relating to climate change, reducing carbon emissions, access to housing and improving the supply of housing, enhancing biodiversity and a number of other emerging priorities. Typically these are addressed via the process of formulating local planning policies for the area of each LPA (Local Planning Authority) on a local basis. Any significant development will also require a variety of different consents from different agencies before commencement, such as approval of construction materials and methods under the relevant Building Regulations). Based on the fundamentals of the Town and Country Planning Act (1947) and various subsequent amendments, over 400 LPA's regulates land use and new buildings relying on a 'plan-led system' whereby development plans are formed and the public is consulted. Subsequent development requires planning permission, which is granted or refused with reference to the development plan as a material consideration.

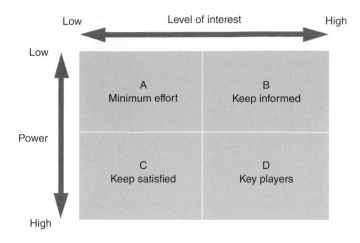

Figure 2.6 Stakeholder mapping: the power/interest matrix. Adapted from Johnson et al. (2006).

Stakeholder identification

Stakeholders are those individuals or groups who directly or indirectly has an influence on and or is influenced by the proposed project.

External stakeholders would usually include financial institutions, customers, end users and members of the public. In some instances, some stakeholders would be identified through regulatory or legislative requirements.

Internal stakeholders will include organisations involved directly in the project which may include suppliers and service providers, and the individuals who have sufficient power to determine the strategy of the organisation.

Stakeholder mapping is the process which identifies stakeholders and the associated expectations and power. It also helps to understand any underlying political priorities within the stakeholders. It underlines the importance of two issues:

- How interested each stakeholder/group is to impress its expectations on the project

- Whether the stakeholders have the power to do so

Figure 2.6 identifies the stakeholder mapping process through the power/interest matrix.

Business case

A business case is used to obtain management commitment and approval for funding or investment through demonstrating a rationale for the funding or investment. The business case provides a framework for planning and management of the proposed development as well as a monitoring benchmark for the ongoing viability of a project.

A template is provided next for the development of a business case in Briefing Note 2.05.

Approval to proceed

The client, reviewing the documents generated throughout this phase, has to re-affirm the decision to proceed with the project, in order to

- Provide the authorisation of financial management and control throughout the project

- Ensure that no commitment is made to large expenditure on the project before verifying that it is sensible to do so

Table 2.2 Client's decision prompt list

- Is there adequate funding for the project?
- Does the project brief demonstrate the existence of a worthwhile project, and hence justify the investment involved?
- Are external support and facility requirements defined and available?
- Have the most appropriate performance standards been applied in order to achieve the best value for money?
- Are assurance responsibilities allocated and accepted?
- Have the agreed sustainable criteria been applied?

At the early stages of a project, it is unlikely that the actual costs will be known. It is important to check the need for financial provision (including time, cost and risk contingencies) has been recognised by all the parties. The predicted project cost at later stages, which will in most cases be different to the original estimate, requires the question of affordability to be revisited at that stage to be sure that adequate funds are available (Table 2.2).

BIM brief

Once the project brief has been established, this will set the parameters for the project. This will include programme, budget, and requirements around function, quality, standard of provision, environmental performance and lifecycle.

Using a BIM approach, it is possible to quite quickly develop basic conceptual options, not only for evaluation by the client but also discussion with other stakeholders such as local community, development control and end users.

In an established BIM environment with experienced practitioners, libraries of data and content will be available. These could include benchmarking data for costs, procurement, product libraries and information from related or similar projects. All of this will help inform the feasibility process, the aim of which is to identify the preferred option to take the project forward. There are also benefits of BIM from a lifecycle perspective in terms of building services, management and replacement of components.

Simple visualisations could be used at this stage to illustrate proposals and simulation and validation techniques used to quantify costs, assess environmental factors, orientation and access requirements for instance. In this way, the project brief can developed in detail and expectations and understanding aligned across the team.

Briefing Note 2.01 Key sustainability issues

Sustainability issues – design and construction objectives (Source: *CIBSE Introduction to Sustainability*. © April 2007 The Chartered Institution of Building Services Engineers, London)

Sustainability issue	Examples of design and construction objectives
Energy and carbon dioxide	Reduce predicted carbon dioxide emissions by applying energy-efficient design principles and using low and zero-carbon technologies
Water	Reduce predicted water use by integrating water-efficient plant, appliances and fittings
Waste	Reduce construction and demolition waste going to landfill and enable in-use recycling in accordance with the waste hierarchy
Transport	Increase the use of sustainable modes of transport when the building is in use
Adapting to climate change	Improve the capacity of the building to operate successfully under the different and demanding conditions predicted in future
Flood risk	Mitigate the risk of flooding (and design for flood resilience)
Materials and equipment	Reduce the embodied lifetime environmental impact by selecting on the basis of environmental preference
Pollution	Reduce unavoidable building-related emissions and the risk of accidental pollution
Ecology and biodiversity	Enhance the ecology and biodiversity of the site by protecting existing assets and by introducing new habitats and/or species
Health and well-being	Provide a safer, more accessible, healthy and comfortable environment
Social issues	Reduce potential for crime and adverse effects on neighbours throughout the lifetime of the development through design and good practice in construction and operation, for example, by adopting secure by design or similar approaches.

Key actions at each project stage (source: *CIBSE Introduction to Sustainability*. © April 2007 The Chartered Institution of Building Services Engineers, London)

Key stage	Key actions
Pre-inception	Identify all drivers for sustainability and ensure that appointment allows for project team to respond to these drivers. Identify the risks associated with project which relate to sustainability (e.g. flood risk assessment, damage to ecological habitat, transport impacts, etc.). Determine potential impact of sustainability targets (e.g. a target for a 'zero-carbon development' is likely to have implications on whole project team).

(Continued)

Key stage	Key actions
	Include scope and fees for early-stage predictions of energy and water use in scope of work (early-stage energy/carbon assessments are becoming essential). Determine whether an environmental impact assessment is required.
Strategic brief	Provide a response to the strategic brief by considering drivers for sustainability and raising issues early in the project. Identify any requirements in the brief that could conflict with sustainability objectives (e.g. design targets for low internal temperatures in summer). Identify requirements for input from specialist consultants (e.g. likely ground conditions for ground source heat pumps).
Project brief	Propose sustainability objectives and targets, in particular carbon and water targets in response to drivers for sustainability. Determine whether assessment methodologies are required (e.g. BREEAM, NEAT) and ensure that project contributes towards all relevant targets. Ensure that design responsibilities are allocated for all critical sustainability targets, especially those relating to carbon and water use.
Strategy	Undertake an initial site analysis against sustainability targets, including determining infrastructure capacity, establishing ground conditions, etc. Provide rules of thumb and design guidance for project team on key issues (e.g. number of wind turbines required to meet predicted loads, or likely spaces for an energy centre). Develop an energy and carbon emissions strategy by following the principles set out in CIBSE *Guide L: Sustainability*. Develop a water management strategy by following the principles set out in CIBSE *Guide L: Sustainability*. Develop a strategy for adapting to the effects of climate change by following the principles set out in CIBSE *Guide L: Sustainability*. Recommend that the project team establishes the flood risk of the site and consults with local authority to determine whether a strategic flood-risk assessment has been undertaken. Incorporate flood-resistant principles into the design of building services and work with the design team to raise awareness of flood risk and flood resistance. Recommend that project teams give consideration to the incorporation of sustainable drainage systems and the potential to integrate with rainwater collection. The project team should liaise with transport planners in order to identify the scope of transportation work required by the local authority. Recommend that a suitably qualified ecologist be involved to undertake and ecological appraisal of the site. Inform the project team of shading benefits of vegetation integrated into the building design and landscape (e.g. green roofs or walls). Incorporate access and inclusion measures identified in the accessibility audit. Recommend that a waste management strategy be prepared for the operation of the building. Consider potential for energy from waste systems. Establish the need for and feasibility of waste management facilities such as compactors, serviced storage spaces, etc.

Key stage	Key actions
	Recommend that the lifecycle impacts of materials and equipment are considered by the project team and that these are considered during the selection of construction methods in terms of ventilation strategies, appropriate thermal mass, etc.
	Make the project team aware of the principles of designing for deconstruction and consider the whole life of services components for recycling or reuse at the end of their life.
	Recommend that there is active engagement and consultation with the local community.
	Highlight the need for consultation with the local police architectural liaison officers on safety and security ('Secure by Design').
	Determine the planning strategy and establish the information that is required for the submission. In particular, determine whether an energy strategy report and sustainability statement are required for application.
	Contribute towards an environmental impact assessment (if required), particularly in relation to air quality, noise, microclimate issues, etc.
Design	Identify the options for reducing demand, supplying efficiently and for providing low- or zero-carbon technologies.
	Propose feasible technologies and techniques to meet carbon emissions targets.
	Identify the options for reducing water demand, supplying water efficiently and for use of rainwater or treatment and reuse of water.
	Propose feasible technologies and techniques to meet water-use targets.
	Advise clients on the maintenance and operational implications associated with using F-Gas refrigerants, such as R134a and R407c.5.3.
	Ensure that the proposals provide comfortable and appropriate internal conditions that promote health and well-being, as set out in the relevant guidance.
	Ensure that storage space for efficient management of waste and recyclable material during operation is incorporated into the layout and that this space is correctly serviced and managed.
	Select and source materials based on the overall environmental impacts and suppliers' declarations.
	Avoid use of environmentally hazardous materials such as insulants with gases implicated in global warming.
	Avoid selecting or locating plant that may create additional noise over the existing background level.
	Consider alternative arrangements for providing and maintaining infrastructure and delivering services, such as energy service companies and multi-utility joint ventures.
	Incorporate all technologies and techniques, as identified in the earlier design stages and refer to the CIBSE online sustainable engineering tool to identify detailed measures.
Construction	Recommend that contractor selection takes account of environmental credentials.
	All relevant tender packages should be reviewed against the sustainability requirements for the project.
	Recommend that subcontractor and supplier selection takes account of environmental credentials.
	Recommend a periodic review of sustainability performance against objectives and targets.

(Continued)

2.01 Briefing Note

Key stage	Key actions
	Ensure that the engineering services that are procured and delivered to site meet the performance standards relating to sustainability, and that the requirements are fully addressed. Observe construction site practices and comment on practices that could have a significant impact on the environment.
Commissioning	Ensure that systems building commissioning/recommissioning results accord with sustainability targets and that the contractor is notified of any issues with performance.
Building handover	Provide a building logbook and occupant user guide for projects and ensure that there is a clear explanation of design targets and assumptions to allow comparison with actual, operational energy use.
Operation	Ensure that the system is operating according to design intent, which may involve periodic recommissioning and post-occupancy evaluation. Recommend that sustainability be addressed when the building owner or occupier specifies tenders and evaluates contracts for the operation and maintenance of facilities. Ensure that refurbishment or refit projects implement the relevant sustainability principles, as set out in this document. Undertake energy and water management activities including audits and benchmarking to identify potential for further savings. Recommend that projects to re-engineer systems consider potential reuse of materials or systems. Ensure that audits and condition surveys include assessment against key sustainability drivers and targets, as identified in this document. Refer to the CIBSE online sustainable engineering tool to identify detailed measures for improving performance.
Deconstruction	An audit should be undertaken prior to demolition commencing to identify the potential for cost-effective recovery of material from demolition.

Briefing Note 2.02   Environmental sustainability assessment methods

BREEAM

The Building Research Establishment Assessment Method (BREEAM 2008) is the leading and most widely used environmental assessment method for buildings (other than homes and dwellings). It sets the standard for best practice is sustainable design and grades performance as pass, good, very good, excellent and outstanding. The assessment of the environmental impacts is carried out at

- Design stage: leading to an interim BREEAM certificate.

- Post-construction stage leading to a final BREEAM certificate and is measured against the following 10 categories:

  - Management
  - health and well-being
  - energy
  - transport
  - water
  - material
  - waste
  - land use and ecology
  - pollution
  - innovation

The following table summarises the main issue in each of the BREEAM categories.

Categories	Issues
Management	Commissioning
	Construction site impacts
	Security
Health and well-being	Daylight
	Occupant thermal comfort
	Acoustics
	Indoor air and water quality
	Lighting
Energy	Carbon dioxide emissions
	Low- or zero-carbon technologies
	Energy sub-metering
	Energy efficient building systems

(Continued)

Categories	Issues
Transport	Public transport network connectivity
	Pedestrian and cyclist facilities
	Access to amenities
	Travel plans and information
Water	Water consumption
	Leak detection
	Water reuse and recycling
Materials	Embodied lifecycle impact of materials
	Materials reuse
	Responsible sourcing
	Robustness
Waste	Construction waste
	Recycled aggregates
	Recycling facilities
Land use and ecology	Site selection
	Protection of ecological features
	Mitigation/enhancement of ecological value
Pollution	Refrigerant use and leakage
	Flood risk
	Nitrous oxide emissions
	Watercourse pollution
	External light and noise pollution
Innovation	Exemplary performance levels
	Use of BREEAM or similar methodology accredited professionals
	New technologies and building processes

Home and dwellings

Code for sustainable homes

For homes and dwellings there are two assessment methods[1]:

- *Code for Sustainable Homes*, April 2007 (new housing in England) issued by the Department for Communities and Local Government.

- The Building Research Establishment's *EcoHomes*, 2006 (existing homes in England and all homes in Scotland, Wales and Northern Ireland). This assessment method formed the basis for the above code.

The *Code for Sustainable Homes* measures sustainability against nine design categories:

- energy and carbon dioxide emissions

- water

- materials

- surface water run-off

- waste

- pollution

---

[1] A major consultation and review is currently being undertaken by the Government, the Housing Standards Review, the outcome of which is likely to have an impact on the building regulation framework and the voluntary housing standards.

- health and well-being

- management

- ecology

The following table summarises the main issues in each of the code categories:

Categories	Issues
Energy and carbon dioxide emissions	Dwelling emission rate (*M*) Building fabric Internal lighting
	Drying space Energy labelled white goods External lighting Low- or zero-carbon technologies Cycle storage Smart working
Water	Internal water use (*M*) External water use
Materials	Environmental impact of materials Responsible sourcing of materials – building elements Responsible sourcing of materials – finishing elements
Surface water run-off	Management of surface water run-off from developments (*M*) Flood risk
Waste	Storage of non-recyclable waste and recyclable household waste (*M*) Construction waste management (*M*) Composting
Pollution	Global warming potential of insulants Nitrous oxide emissions
Health and well-being	Daylighting Sound insulation Private space Lifetime homes (*M*)
Management	Home use guide Considerate constructors scheme Construction site impacts Security
Ecology	Ecological value of site Ecological enhancement Protection of ecological features Change in ecological value of site Building footprint

(*M*) *denotes mandatory elements.*

The code uses a rating system of one to six stars with one star being the entry level just above the Building Regulations and six stars being evidence of exemplar sustainable development. Credits are given for achieving a certain level of performance in each category and the factored against a weighting system see the following figures.

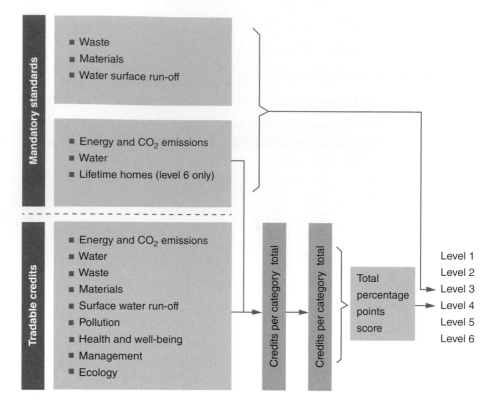

Scoring system for the *Code for Sustainable Homes*.

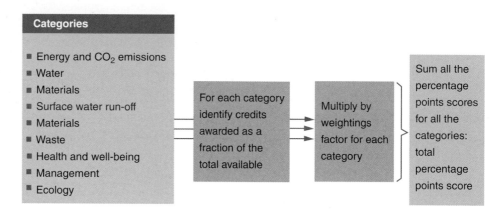

Calculating the total points score.

As with BREEAM, the assessment is carried out at the design and post-construction stages.

EcoHomes

EcoHomes measures sustainability against eight similar design categories:

- energy
- transport
- pollution
- materials

- water

- land use and ecology

- health and well-being

- management

and employs a similar credit rating system which classifies the development under pass, good, very good and excellent. The following table summarises the main issues in each EcoHomes category.

Categories	Issues
Energy	Dwelling emission rate Building fabric Drying space
	EcoLabelled goods Internal lighting External lighting
Transport	Public transport Cycle storage Local amenities Smart working
Pollution	Insulant global warming potential Nitrous oxide emissions Reduction and surface run-off Renewable and low emission energy source
Materials	Flood risk Environmental impact of materials
	Responsible sourcing of materials: basic building elements Responsible sourcing of materials: finishing elements Recycling facilities
Water	Internal potable water use External potable water use
Land use and ecology	Ecological value of site Ecological enhancement Protection of ecological features Change of ecological value of site Building footprint
Health and well-being	Daylighting Sound insulation Private space
Management	Home use guide Considerate constructors Construction site impacts

2.02 Briefing Note

Briefing Note 2.03   Guidance on environmental impact assessment

Introduction

Environmental impact assessment (EIA) is a key instrument of EU environmental policy. Since passage of the first EIA directive in 1985 (Directive 85/337/EEC) both the law and the practice of EIA have evolved. An amending directive was published in 1997 (Directive 97/11/EC), further amended in 2003 (Directive 2003/35/EC).

Assessment of the effects of certain public and private projects on the environment is required under the Town and Country Planning impact (Environmental impact Assessment) (England and Wales) Regulations 1999 and subsequent amendments, in so far as it applies to development under the Town and Country Planning Act 1990. EIA is a means of drawing together, in a systematic way, an assessment of a project's likely significant environmental effects. This helps to ensure that the importance of the predicated effects, and the scope for reducing them, are properly understood by the public and the relevant competent authority before it makes its decision.

Where an EIA is required there are three broad stages to the procedure:

- The developer must compile detailed information about the likely main environmental effects. To help the developer, public authorities must make available any relevant environmental information in their possession when requested. The developer can also ask the 'competent authority' for its opinion on what information needs to be included. The information finally compiled by the developer is known as an environmental statement (ES).

- The ES (and the application to which it relates) must be publicised. Public authorities with relevant environmental responsibilities and the public must be given an opportunity to give their views about the project and ES.

- The ES, together with any other information, comments and representations made on it, must be taken into account by the competent authority in deciding whether or not to give consent for the development. The public must be informed of the decision and the main reason for it.

The regulations

The regulations integrate the EIA procedures into the existing framework of local authority control. These procedures provide a more systematic method of assessing the environmental implications of developments that are likely to have significant effects. EIA is not discretionary. If significant effects on the environment are likely, EIA is required. Where the EIA procedure reveals that a project will have an adverse impact on the environment, it does not follow that planning permissions must be refused. It remains the task of the local planning authority to judge each planning application on its merits within the context of the development plan, taking account of all materials considerations, including the environmental impacts.

For developers, EIA can help to identify the likely effects of a particular project at an early stage. This can produce improvements in the planning and design of the development, in decision making by both parties, and in consultation and responses thereto, particularly if combined with early consultations with the local planning authority and other interested bodies during the preparatory stages. In addition, developers may find EIA a useful tool for considering alternative approaches to a development. This can result in a final proposal that is more environmentally acceptable, and can form the basis for a more robust application for planning permission. The presentation of environmental information in a more systematic way may also simplify the local planning authority's task of appraising the application and drawing up appropriate planning conditions, enabling swifter decisions to be reached.

For EIA applications, the period after which an appeal against non-determination may be made is extended to 16 weeks.

Environmental impact assessment (EU regulations)

EIA is a procedure required under the terms of EU directives on assessment of the effects of certain public and private projects on the environment.

Key stages	Notes
Project preparation	The client prepares the proposals for the project.
Notification to competent authority	The competent authority may be the local authority, Environment Agency, English Nature or a similar organisation depending on the nature and the location of the project.
Screening	The competent authority makes a decision on whether EIA is required. This may happen when the competent authority receives notification of the intention to make a development consent application or the developer may make an application for a screening opinion. The screening decision must be recorded and made public.
Scoping	The EU directive provides that developers may request a scoping opinion from the competent authority. The scoping opinion will identify the matters to be covered in the environmental information. It may also cover other aspects of the EIA process.
Environmental studies	The developer carries out studies to collect and prepare the environmental information required.
Submission of the environmental statement to Competent Authority	The developer submits the environmental information to the competent authority together with the application for development consent. The environmental information is presented usually in the form of an environmental impact statement.
Review of adequacy of the environmental information	The client may be required to provide further information if the submitted information is deemed to be inadequate.
Consultation with statutory environmental authorities, other interested parties and the public	The environmental information must be made available to authorities with environmental responsibilities and to other interested organisations and the general public for review.
	They must be given an opportunity to comment on the project and its environmental effects before a decision is made on development consent.

(Continued)

Key stages	Notes
Consideration of the environmental information by the competent authority before making development consent decision	The environmental information and the results of consultations must be considered by the competent authority in reaching its decision on the application for development consent.
Announcement of decision	The decision must be made available to the public including the reasons for it and a description of the measures that will be required to mitigate adverse environmental effects.
Post-decision monitoring if project is granted consent	There may be a requirement to monitor the effects of the project once it is implemented.

Establishing whether EIA is required

Generally, it will fall to local planning authorities in the first instance to consider whether a proposed development requires EIA. For this purpose, they will first need to consider whether the development is decried in Schedule 1 or Schedule 2 to the regulations. Development of a type listed in Schedule 1 always requires EIA. Development listed in Schedule 2 requires EIA if it is likely to have significant effects on the environment by virtue of factors such as its size, nature or location.

Development which comprises a change or extension requires EIA only if the change or extension is likely to have significant environmental effects.

Like the Town and Country Planning Act, the Regulations do not bind developments by Crown bodies.

Planning applications

The requirements for EIA for planning applications must be established in accordance with the Town and Country Planning (Environmental Impact Assessment) Regulations 2011 and guidance in DETR circular 2/99. When any planning application is made in outline, the local planning authority will need to satisfy itself that it has sufficient information available on the environmental effects of the proposal to determine whether or not planning permission should be granted in principle.

Where the authority's opinion is that EIA is required, but not submitted with the planning application, it must notify the applicant within three weeks of the date of receipt of the application, giving full reasons for its view clearly and precisely.

An applicant who still wishes to continue with the application must reply within three weeks of the date of such notification. The reply should indicate the applicant's intention either to provide an environmental statement or to ask the secretary of state for a screening direction. If the applicant does not reply within three weeks, the application will be deemed to have been refused.

Preparation and content of an environmental statement

It is the applicant's responsibility to prepare the ES. There is no statutory provision as to the form of an ES. However, it must contain the information specified in Part II, and such of the relevant information in Part I of Schedule 4 to the regulations as is reasonably required to assess the effects of the project and which the developer can reasonably be required to compile.

The list of aspects of the environment which might be significant affected by a project is set out in paragraph 3 of Part I of Schedule 4, and includes human beings, flora, fauna, soil, water, air, climate, landscape, material, assets, including architectural and archaeological heritage and the interaction between any of the foregoing.

Procedures for establishing whether or not EIA is required ('screening')

The determination of whether EIA is required for a particular development proposal can take place at a number of different stages:

- The developer may decide that EIA will be required and submit a statement.

- The developer may, before submitting any planning application, request a screening opinion from the local planning authority. If the developer disputes the need for EIA (or a screening opinion is not adopted within the required period), the developers may apply to the secretary of state for a screening direction. Similar procedures apply to permitted development.

- The local planning authority may determine that EIA is required following receipt of a planning application. If the developer disputes the need for EIA, the applicant may apply to the secretary of state for a screening direction.

- The secretary of state may determine that EIA is required for an application that has been called in for his determination or is before him on appeal.

- The secretary of state may direct that EIA is required at any stage prior to the granting of consent for particular development.

Provision to seek a formal opinion from the local planning authority on the scope of an ES ('scoping')

Before making a planning application, a developer may ask the local planning authority for their formal opinion on the information to be supplied in the ES (a 'scoping opinion'). This provision allows the developer to be clear about what the local planning authority considers the main effects of the development are likely to be and, therefore, the topics on which the ES should focus. The developer must include the same information as would be required to accompany a request for a screening opinion. The local planning authority must adopt a scoping opinion within five weeks of receiving a request.

Provision of information by the consultation bodies

Under the Environmental Information Regulations, public bodies must make environmental information available to any person who requests it. Once a developer has given the local planning authority notice in writing that he intends to submit an ES, the authority must inform the consultation bodies. The consultation bodies are

- The bodies who would be statutory consultees under article 10 of the Development Management Procedure Order (2010 Article 16 Appendix 5) for any planning application for the proposed development.

- Any principal council for the area in which the land is situated (other than the local planning authority)

- Nature English (NE)

- Scottish Natural Heritage

- Natural Resources Wales (NRW)
- Department of Environment Northern Ireland (DENI)
- Joint Nature Conservation Committee
- Council for Nature Conservation and the Countryside
- Scottish Environment Protection Agency
- The Environment Agency
- English Heritage

The characteristics of a good environmental statement

- A clear structure[2] with a logical sequence, for example, describing, existing baseline conditions, predicted impacts (nature, extent and magnitude), scope for mitigation, agreed mitigation measures, significance of unavoidable/residual impacts for each environmental topic.

- A table of contents at the beginning of the document.

- A clear description of the development consent procedure and how EIA fits within it.

- Reads as a single document with appropriate cross-referencing.

- Is concise, comprehensive and objective.

- Written in an impartial manner without bias.

- Includes a full description of the development proposals.

- Makes effective use of diagrams, illustrations, photographs and other graphics to support the text.

- Uses consistent terminology with a glossary.

- References all information sources used.

- Has a clear explanation of complex issues.

- Contains a good description of the methods used for the studies of each environmental topic.

- Covers each environmental topic in a way which is proportionate to its importance.

- Provides evidence of good consultations.

- Includes a clear discussion of alternatives.

- Makes a commitment to mitigation (with a programme) and to monitoring.

- Has a non-technical summary which does not contain technical jargon.

---

[2] Schedule 4 of the Regulation.

1.0   Introduction

1.1   General project description
1.2   EIA development
1.3   Planning context
1.4   Scope and content of the ES
1.5   ES availability and comments

2.0   EIA Methodology

2.1   Objectives
2.2   Scoping study
2.3   Consultations
2.4   Defining the baseline
2.5   Sensitive receptions
2.6   Impact prediction
2.7   Evaluation of significance
2.8   Mitigation
2.9   Residual impact
2.10 Assumptions and limitations

3.0   Development Background and Alternatives

3.1   Introduction
3.2   Site considerations and constraints
3.3   No development alternative
3.4   Objectives of the proposed redevelopment
3.5   Design alternatives

4.0   The Site Description and Design Statement

4.1   Introduction
4.2   Site location and setting
4.3   Site description
4.4   The design statement

5.0   Plant Dismantling, Demoltion, Remediation and Construction

5.1   Introduction
5.2   Schedule overview
5.3   Plant dismantling and asbestos removal
5.4   Demolition
5.5   Remediation
5.6   Construction
5.7   Construction traffic
5.8   Environmental management plan and code of construction

6.0   Environmental Management Plan (EMP) and Potential Impacts of the Construction Works

6.1   Introduction
6.2   Scope of the EMP
6.3   Summary and conclusions

7.0   Planning and Policy Context

7.1   Introduction
7.2   Planning policy guidance
7.3   Strategic guidance
7.4   Strategic planning in (location)
7.5   Planning brief
7.6   The adopted local UDP
7.7   UDP proposed alterations
7.8   Affordable and social housing
7.9   Summary and conclusions to planning and policy context

8.0   Sustainability – Environmental

8.1   Introduction
8.2   National guidance and local policy
8.3   Approach to assessment
8.4   Sustainability topics
8.5   Results
8.6   Summary and conclusions

9.0   Socio-economic impacts

9.1   Introduction
9.2   Approach to assessment
9.3   Baseline statistics
9.4   Impact of the development
9.5   Summary and conclusions

10.0   Built Heritage, Townscape and Visual Impacts

10.1   Introduction
10.2   Approach to assessment
10.3   Approach to presentation of the visual assessment
10.4   Baseline condition – the heritage and existing townscape
10.5   Townscape studies, policies and guidelines
10.6   Townscape and visual impact assessment
10.7   Impact assessment – the immediate locality
10.8   Impact assessment – the panorama, the moving eye and the interaction of major built forms
10.9   Impact assessment – specific viewpoints reviewed
10.10 Summary and conclusions

11.0   Archaeology

11.1   Introduction
11.2   Approach to assessment
11.3   Policy considerations and legislative
11.4   Initial assessment
11.5   Archaeological potential of the site
11.6   Environmental potential of the site
11.7   Archaeological resources in the surrounding area
11.8   Summary of archaeological potential
11.9   Impact of the development
11.10 Mitigation
11.11 Summary and conclusions

2.03 Briefing Note

2.03 Briefing Note

12.0 Water Resources

12.1 Introduction
12.2 Approach to assessment
12.3 Methodology and assumptions
12.4 Baseline groundwater conditions
12.5 Baseline surface water resources
12.6 Amenity/recreation
12.7 Impact of the no development alternative on (location)
12.8 Impact of the development option on (location)
12.9 Mitigation and proposed monitoring
12.10 Residual effects
12.11 Summary and conclusions

13.0 Ecology

13.1 Introduction and background
13.2 The ecological assessment
13.3 Methodology
13.4 Existing baseline
13.5 Baseline ecological evaluation
13.6 Predicted changes to existing baseline – the no development alternative
13.7 Potential impacts of the development proposals and mitigation proposals
13.8 Design responses to ecological issues
13.9 Summary and conclusions

14.0 Contamination Issues

14.1 Introduction
14.2 Approach to assessment
14.3 Site history
14.4 Above ground survey results
14.5 Below ground survey results
14.6 Risk assessment
14.7 Remediation design
14.8 Impacts of the development and mitigation
14.9 Summary and conclusions

15.0 Transport

15.1 Introduction
15.2 Approach to assessment
15.3 Overview of existing public transport network
15.4 The existing highway network
15.5 Predicted traffic flows and assessment of impact of the development
15.6 Transport initiatives and mitigation
15.7 Summary and conclusions

16.0 Air Quality

16.1 Introduction
16.2 Current legislation and air quality standards
16.3 Planning policy context
16.4 Approach to assessment
16.5 Baseline information
16.6 Mitigation measures
16.7 Air quality modelling study
16.8 Assessment of impacts
16.9 Summary and conditions

17.0 Noise and Vibration

17.1 Introduction
17.2 Approach to assessment
17.3 Assessment methodology
17.4 Identification of sensitive receptors
17.5 Baseline environment
17.6 Estimation of the noise and vibration levels generated during redevelopment
17.7 Estimation of the noise and vibration levels generated during operation
17.8 Impact assessment of the redevelopment
17.9 Impact assessment on new occupied dwellings on the site
17.10 Mitigation
17.11 Impact assessment on new occupied dwellings on the site
17.12 Summary and conclusions

18.0 Micro Climate

18.1 Introduction
18.2 Approach to assessment
18.3 Results
18.4 Impacts on the development without mitigation
18.5 Mitigation
18.6 Summary and conclusions

19.0 Telecommunications

19.1 Introduction
19.2 Approach to assessment
19.3 Methodology
19.4 Survey results
19.5 Potential impacts on the development
19.6 Mitigation
19.7 Summary and conclusions

20.0 Summary of Residual Impacts

20.1 Introduction
20.2 Assessment method
20.3 Conclusions

21.0 Glossary

22.0 Abbreviations

23.0 References

Briefing Note 2.04　Site investigation

This flow chart is used for each of the 10 activities identified in the following table.

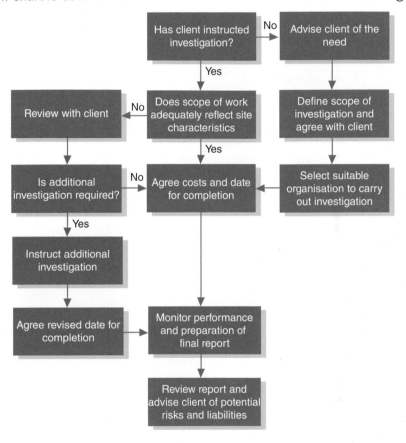

Site investigation activities.

Activities associated with site investigation

Activity	Action by
Site surveys	Land surveyor and structural engineer
Geotechnical investigation	Ground investigation specialist
Drainage and utilities survey	Civil engineering consultant
Contamination survey	Environmental and/or soil specialist
Traffic study	Transportation consultant
Adjacent property survey	Buildings/party walls/rights of light surveyors
Archaeological survey	Local museum or British Museum and other relevant sources
Sustainability issues	Specialist consultant
Legal aspects	Solicitor
Outline planning permission	Architect

Confirmation that the activities have been successfully completed is the responsibility of the project manager.

Each task can be broken down into a number of specific elements.

Site surveys

- location
- ordnance survey reference
- ground levels/contours
- physical features (e.g. roads, railways, rivers, ditches, trees, pylons, buildings, old foundations, erosion)
- existing boundaries
- adjacent properties
- site access
- structural survey
- previous use of site

Geotechnical investigation

- trial pits
- boreholes and borehole logs
- geology of site including underground workings
- laboratory soil tests
- site tests
- groundwater observation and pumping tests
- geophysical survey

Drainage and utilities survey

- existing site drainage (open ditch, culvert or piped system)
- extent of existing utilities on or nearest to the site (water, gas, electricity, telecoms)
- extent of any other services that may cross the site (e.g. telephone/data lines, oil/fuel pipelines)

Contamination survey

- asbestos
- methane
- toxic waste
- chemical tests
- radioactive substances

Traffic survey

- examination of traffic records from local authority
- traffic counts
- traffic patterns
- computer simulation of existing traffic flows

- delay analysis
- noise levels

Adjacent property survey

Traffic survey

- right of light
- party-wall agreements
- schedule of conditions
- foundations
- drainage
- access
- public utilities serving the property
- noise levels (e.g. airports, motorways, air-conditioning equipment)

Archaeological survey

- examination of records
- archaeological remains.

Sustainability issues

- effects of proposed development on local environment
- environmental impact assessment
- flood risk
- carbon dioxide emissions
- waste
- transport
- pollution
- ecology and biodiversity
- health and well-being
- social issues

Legal aspects

- ownership of site
- restrictive covenants
- easements, for example, rights of way, rights of light
- way-leaves
- boundaries
- party-wall agreements
- highways agreements
- local authority agreements
- air rights

Outline planning permission

- effect of local area plan

2.04 Briefing Note

Briefing Note 2.05   Business case development

A business case is used to obtain management commitment and approval for funding or investment through demonstrating a rationale for the funding or investment. The business case provides a framework for planning and management of the proposed development as well as a monitoring benchmark for the ongoing viability of a project.

A template is provided next for the development of a business case.

1. Definition of the project proposal

2. Objective of the project proposal

3. Strategic fit.

   3.1   Business need

   3.2   Organisational overview

   3.3   Contribution to key organisational objectives

   3.4   Stakeholders

   3.5   Existing arrangements

   3.6   Scope (minimum, desirable and optional)

   3.7   Constraints

   3.8   Dependencies

   3.9   Strategic benefits

   3.10 Strategic risks

   3.11 Critical success factors

4. Options appraisal

   4.1   Long and short list of options

   4.2   Opportunities for innovation and collaboration

   4.3   Service delivery options – who will deliver the project?

   4.4   Environmental, social and economic criteria

   4.5   Implementation options

   4.6   Detailed options appraisal demonstrating value for money and sustainability

   4.7   Risk quantification and sensitivity analysis

   4.8   Benefits appraisal

   4.9   Preferred option

5. Commercial aspects

   5.1   Output-based specification

   5.2   Sourcing options

   5.3   Risk allocation and transfer

   5.4   Contract length

   5.5   Implementation timescales

6. Affordability

6.1 Budgetary issues

6.2 Income and expenditure

6.3 Cashflow prediction

7. Achievability

7.1 Evidence of similar projects

7.2 Project roles

7.3 Delivery strategy

7.4 Risk management strategy

7.5 Benefits realisation plan

7.6 Contingency plan

The effectiveness of a business plan should be judged on the basis of the following key criteria:

- Is the need for the project clearly stated?
- Have the benefits been clearly identified?
- Are the reasons for and benefits of the project consistent with the overall strategy?
- Is it clear what will define a successful outcome?
- Is it clear what the preferred option is?
- Is it clear why this is the preferred option?
- Is it clear what the procurement option is?
- Is it clear why this is the preferred procurement option?
- Is it clear how the necessary funding will be put in place?
- Is it clear how the benefits will be realised?
- Are the risks faced by the project explicitly stated?
- Are the plans for addressing those risks explicitly stated?

2.05 Briefing Note

3 Strategy

Stage checklist

Key processes:	Project governance parameters
	Project strategy
	Project organisation and control
	Accountability and responsibility
	Selection and appointment of project team
	Procurement strategy
	Tender procedures
	BIM strategy
	Project execution plan
Key objective:	'How will the need be realised?'
Key deliverables:	Project execution plan
Key resources:	Client team
	Project manager
	Specialist consultants

Stage process and outcomes

This project manager–led stage takes the client's preferred project option which received approval at the completion of the feasibility stage and develops firm proposals defining how the development will be implemented. Determining how the project will be implemented and managed is a fundamental element of the project management role.

This process involves establishing the project infrastructure in terms of defining the mechanisms for managing and controlling the project, assembling the team who will execute the project, and collating all information to clarify the client's detailed requirements as a precursor to the design process.

Essential to this stage is the production of a project execution plan (or project management plan) which confirms the project strategy and the manner in which the project will be implemented.

Outcomes:

- project implementation strategy
- project definitions and procedures
- selection and appointment of project team
- employer's requirement document
- facility management strategy

Code of Practice for Project Management for Construction and Development, Fifth Edition. Chartered Institute of Building.
© Chartered Institute of Building 2014. Published 2014 by John Wiley & Sons, Ltd.

- project time schedule
- project capital budget
- project risk register
- health & safety plan & file
- procurement strategy
- consultants scope of services
- project execution/management plan (PEP/PMP)
- client approval to PEP/PMP
- client approval to proceed to the pre-construction stage

Client's objectives

The aims for the client at this stage include setting up the project organisation, establishing the strategies for procurement, delivery (cost, time and quality control and risk management) and commissioning/occupation issues through identifying project targets, assessing and managing risks and establishing the project plan (Figure 3.1 and Figure 3.2).

Project governance

Project governance helps to ensure that a project is executed according to the standards of the organisation performing the project. Governance keeps all project activities transparent and ethical, and also creates accountability.

A project governance structure will also help to define a project reporting system. It outlines specific roles and responsibilities for everyone involved in the project. Project managers can leverage a governance structure in their projects to help with setting project priorities.

By understanding how governance fits into the larger organisation, a project manager can choose which objectives to pursue or indeed can gain support to change objectives that do not align with the overall organisational goal. By monitoring governance, the project manager helps to ensure that the project will stay in tune with organisational expectations and remains a good investment as it continues in its lifecycle.

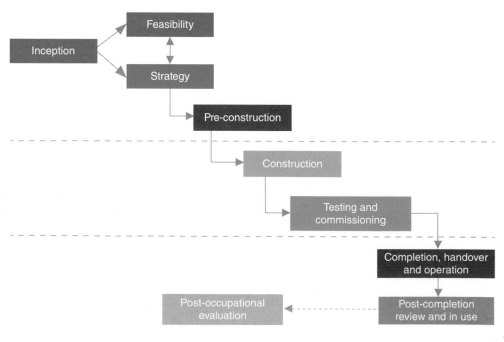

Figure 3.1 Stages of the project development.

3 Strategy

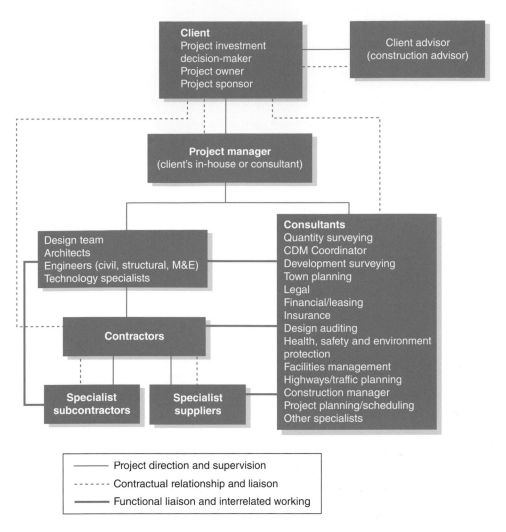

Figure 3.2 Typical project team structure.

A project manager can also use the steering committees or project boards that are part of most governance structures to resolve conflicts. As these high-level committee members do not work on the project on a daily basis, they can serve as fresh eyes to see what is causing the conflict and offer an outside voice of reason. They can also offer solutions on how to resolve the conflict and adhere to the standards – while still sticking to the overall goals of the organisation.

Project governance extends the principle of governance into both the management of individual projects via governance structures, and the management of projects at the business level, for example, via business reviews of projects. Today, many organisations are developing models for 'project governance structures', which can be different to a traditional organisation structure in that it defines accountabilities and responsibilities for strategic decision-making across the project, often utilising the RACI[1] model (**R**esponsible **A**ccountable **C**onsulted **I**nformed). This can be particularly useful to project management processes such as change control and strategic (project) decision making. When implemented well, it can have a significantly positive effect on the quality and speed of decision making on significant issues on projects.

Effective governance should create an environment where no project that is clearly exhibiting commonly accepted characteristics of project failure should be allowed to proceed to their next phase without clear resolution of those issues.

[1]   Mike Jacka, J. & Keller, P.J. (2009) *Business Process Mapping: Improving Customer Satisfaction*. p. 257. John Wiley & Sons, New York.

3 Strategy

It is also worth considering that since corporate governance now places responsibilities on boards to monitor enterprise performance, there is a further responsibility of 'governance of project management'. This encompasses the need to control and demonstrate

- assurance that projects are being managed well and in accordance the requirements of governance across the enterprise

- assurance that portfolio management is optimising the return from corporate resource and maintaining alignment with strategic objectives

- assurance that strategic projects are not exhibiting (well publicised) conditions of project failure

The Association for Project Management outlines 11 principles (see Briefing Note 3.12) for good project management and states that there are four key component of project governance:

- portfolio direction

- project sponsorship

- project management effectiveness and efficiency

- disclosure and reporting

A 2011 publication by the Society for Construction Law[2] mapped the common causes of project failure published in a 2005 report[3] by OGC, against the four components of project governance as set out by the Association for Project Management (see Table 3.1).

Table 3.1 Mapping common causes of project failure

OGC: common causes of project failure	APM: component of project governance
Lack of clear links between the project and the organisation's key strategic priorities, including agreed measures of success	1. Portfolio direction 2. Project sponsorship
Lack of clear senior management and the higher level ownership and leadership	2. Project sponsorship
Lack of effective engagement with stakeholders	2. Project sponsorship 3. Project management
Lack of skills and proven approach to project management and risk management	3. Project management
Lack of understanding of, and contact with, the supply industry at senior levels in the organisation	3. Project management
Too little attention to breaking development and implementation into manageable steps	3. Project management 4. Disclosure and reporting
Evaluation of proposals driven by initial price rather than long-term value of money (especially securing business benefits)	2. Project sponsorship
Lack of understanding of and contract with supply industry at senior levels in the organisation	3. Project management 4. Disclosure and reporting
Lack if effective project team integration between clients, the supplier and the supply chain	3. Project management

Morgan, A. & Gbedemah, S. (2010) How poor project governance causes delays. A paper presented to the *Society of Construction Law at Meeting in London*, 2 February 2010.

[2] Morgan, A. & Gbedemah, S. (2010) How poor project governance causes delays. A paper presented to the *Society of Construction Law at Meeting in London*, 2 February 2010.

[3] Office of Government Commerce (2005) Common causes of project failure, http://www.dfpni.gov.uk/content_-_successful_delivery-newpage-50 (accessed March 2014).

Strategy outline and development

A typical strategy stage consists of the main elements shown in Figure 3.3.

The project manager performs several principal activities at this stage which may include all or most of the following:

- Developing and reviewing the detail project brief with the client and any existing members of the project team to ascertain that the client's objectives will be met. Preparing a final version in written form with supplementary appendices where these add to the general understanding of the issues that support the brief itself.

- Establishing, in consultation with the client and consultants, a project management structure and the participants' roles and responsibilities, including access to client and related communication routes, and 'decision required' points. This should be developed and presented in the project documents for the reference of all parties.

- Ensuring, in liaison with the client, the CDM coordinator, design consultants and the principal contractor when appointed, that appropriate arrangements have been made to meet the requirements of the Construction (Design and Management) (CDM) Regulations. Key duties under the regulations are summarised in Briefing Note 3.01.

- Developing the client's performance brief for environmental sustainability including targets for recognised assessments such as BREEAM, LEED and Energy Performance, determining how this will be assessed and whether a specialist consultant such as environmental performance assessor should be appointed. As part of this, establishing investment criteria for environmental measures and including them in value management reviews.

- Establishing that 'risk and value management' principles are applied effectively from the earliest stages of the preparation of the design brief until the design is complete. The emphasis should be on providing value for money and in producing a facility that can be constructed and operated at the optimum time and cost without compromising quality, scope or specifications. The design team and

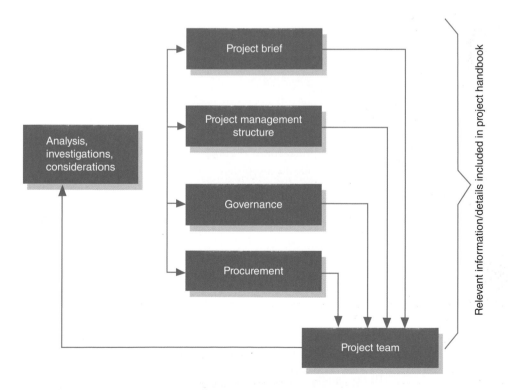

Figure 3.3 Elements of the strategy stage.

3 Strategy

consultants should be encouraged to consciously seek to identify the optimum balance of time and cost by better design and construction logistics. An approach where the whole team 'designs in quality and time-effectiveness and drives out cost' at all stages in the design process should be encouraged – emphasis on overall value should be encouraged. Further guidance on value management is included Briefing Note 3.02.

- Advising the client on the recruitment and appointment of additional consultants and design team members, that is:

  - preparation of appropriate definition of roles and responsibilities

  - preparation and issue of selection/tender documentation

  - evaluation, reporting and making recommendations

  - assisting the client in the preparation of agreements and in selection and appointment

- Drawing the client's attention to the benefits of project insurance for the whole team and works and providing assistance in assessing the risks on the project and including an appropriate contingency sum in the project budget and time contingencies in the master development schedule. Putting in place procedures for managing risk as a continuous project activity. A project risk assessment checklist (Briefing Note 3.03) may be used or adapted as part of such a procedure. (These risks are not to be confused with risks covered by the CDM Regulations, although CDM risks will form a subset of an overall risk management regime.):

- Selection, or development, and agreement of the most appropriate form of contract relative to the project objectives and the parameters of cost, time, quality, function and financial viability and risk management.

- Assisting the client in completing site selection/evaluation, investigation and acquisition.

- Advising on whether certain activities, such as fitting out and occupation/migration, constitute separate projects and should be treated as such.

- Making the client aware of relevant statutory submissions and other consultations that may be required in the delivery of the project.

Project organisation and control

A project management organisation structure sets out unambiguously and in detail how the parties to the project are to perform their functions in relation to each other in contributing to the overall scheme. This should be recorded in the *Project Handbook* (see Glossary for definition and Briefing Note 1.04). It also identifies arrangements and procedures for monitoring and controlling the relevant administrative details. It is updated as circumstances dictate during the lifetime of the project and should allow project objectives and success criteria to be communicated and agreed by all concerned and help promote effective teamwork.

Procedures covering the relationships and arrangements for record keeping, monitoring progress, time and cost control, risk management, project control and administration of the project should be developed, with the assistance of parties involved, for all stages of the project and cover time, costs, quality and reporting/decision-making arrangements.

The organisation structure should clearly identify the involvement and obligations of the client and his organisational backup.

Project team structure

Under the overall direction of a project manager, projects are usually carried out by a project team typically comprising the following:

- client's internal team (appropriate representatives)

- project manager (either within the client's own organisation or independently appointed)

- design team: architects, structural/civil/mechanical and electrical (M&E) engineers and technology specialists

- consultants covering quantity surveying, development surveying, planning and scheduling, legal issues, valuation, finance/leasing, insurances, design audit, sustainability and energy certification, health and safety and environmental protection, access issues, facilities management, highways/traffic planning, construction management, and other specialisms

- contractors, subcontractors and suppliers

The project team structure for project management is shown in Figure 3.2. This structure is idealised and, in practice, there will be many variants depending on the nature of the project, the contractual arrangements, type of project management (external or in-house) involved, and above all, the client's requirements. It should be one of the duties of the project manager to advise the client on the most appropriate project team structure for a particular project.

Effective project management must, at all times, fully embrace all provisions for quality assurance, time and financial control, health and safety, information exchange and environmental protection. These aspects are to be considered as incorporated and implied in all relevant activities presented in this Code of Practice.

Selecting the project team

When establishing a project team, many skills will be needed. During selection, the project manager should consider the following factors:

- A commitment by the project team to clearly defined and measurable project objectives.

- Firm duties of teamwork, with shared financial motivation to pursue those objectives. These should involve a general presumption to achieve 'win-win' solutions to problems which may arise during the course of the project. Issues such as leadership, communication and teamworking form key cornerstones of a successful project delivery.

- The production of satisfactory evidence from each team member, to show that they can contribute effectively to the project objectives. This evidence may include a realistic schedule with appropriate allocation of contingencies against foreseeable risks, a financial plan and a demonstration of adequate resources.

- When choosing each team member, as suggested in the Inception stage, special attention to be paid to their:

  - relevant experience

  - technical qualifications

  - appreciation of project objectives

  - level of available supporting resources

3 Strategy

- creative/innovative ability

- enthusiasm and commitment

- positive team attitude

- communication skills

- Financial strength and core resource strength are also important.

- Defining clear lines of communication between the respective project team members.

- Promoting a working environment that encourages an interchange of ideas by rewarding initiatives which ultimately benefit the project.

- Undertaking regular performance appraisals for all project team members.

- Ensuring that project team members are suitably located and that communication protocols have been established (particularly for electronic sharing of information) so as to facilitate regular contact with each other, as well as with their own organisations.

- Defining clear areas of responsibility and lines of authority for each project team member, and communicating these within the team.

- Identifying a suitable deputy for each team member, who will be sufficiently familiar with the project to be able to act as their replacement should the need arise.

- Making provision for members of the project team to meet informally and socially, outside the work environment, on a regular basis.

Project management procedures and systems

There are two main elements in the management of construction projects. One is the overall management function, which has much in common with the management of any other kind of organisation. It is concerned with the direction and coordination of the decision-making processes within the project area and its purpose is to get all those involved in the process (the stakeholders including the project team members) to work towards the achievement of the project's objectives. The other element is the management of specialised activities which are peculiar to specific projects. These activities have certain characteristics calling for particular kinds of management skills and styles of operation and it is the responsibility of the project manager to identify these specialised activities and ensure that adequate and appropriate processes and procedures are in place to effect efficient management of these activities.

The successful delivery of the project may depend on the project manager identifying and obtaining specialist input at appropriate stages and to the appropriate extent.

Information and communication technology

The construction industry is one of the largest contributors of wealth creation to Europe's business economy, accounting for almost one-tenth of the gross domestic product (GDP) and almost two-thirds of the gross fixed capital formation. The industry is typically extremely information intensive[4] and knowledge based and, therefore, construction organisations in order to be successful and efficient need to fully embrace information and communication technologies (ICT). Although the industry is traditionally renowned for low margins, low levels/barriers to market entry, poor investment in

---

[4] Underwood, J. & Khosrowshahi, F. (2012) ITC expenditure and trends in the UK construction industry in facing the challenges of the global economic crisis. *Journal of Information Technology in Construction (ITcon)*, 17, 25–42, http://www.itcon.org/2012/2 (accessed April 2014).

research and development (R&D), etc., ICT has continued to receive significant levels of growth and integration.[5]

The key guidance and information necessary to strategise and formulate ICT relating to governance and management of projects can be located in Briefing Note 3.04, a summary of which is stated below:

- Business process – Construct IT for Business (www.construct-it.org.uk)

- Interoperability – International Alliance for Interoperability (IAI) Building SMART – UK Chapter (www.buildingsmart.org.uk)

- eBusiness – The Network for Construction Collaboration Technology Providers (NCCTP) (Part of Construction Excellence initiatives)

- Electronic Document and Records Management Systems (EDRMS) – Information and advice available from Construct IT and the IT Construction Forum of Construction Excellence.

- Electronic Trading – Construction Industry Trading Electronically (CITE) – Information available from Construction Excellence website.

- Enterprises Resources Planning (ERP) – ERP traces its roots to the manufacturing sector where Material Requirement Planning (MRP) systems first used computers to automate planning for components. A range of software is available for use in the construction industry.

- Concurrent Engineering – The concept of concurrent engineering originated in the field of product development where multiple functions such as design and manufacturing were integrated to accelerate the delivery of the product.

- Information visualisation – With increased technological advances in computer-generated imagery (CGI) technology and assimilation of processes such as Building Information Modelling (BIM), information visualisation is being more utilised than perhaps ever before.

- Mobile technology – Mobile data management systems are being used to manage several technical functions which include the following:

  - Health and safety

  - Drawing distribution and usage

  - Goods received notes

  - Maintenance and snagging inspections

  - Monitoring hazardous activities

  - Monitoring progress

  - Monitoring resources

  - Quality inspection

  - On-site design issues resolution

  - Site diaries

  - Task allocation and monitoring

In view of the continual growth and popularity of 3D and 4D softwares particularly through BIM, ICT constitutes an important and critical tool in successful development and delivery of construction projects.

[5] 2020 Vision – The Future of UK Construction (2008) – A scenario based report by SAMI Consulting for ConstructionSkills.

Examples and further guidance towards successfully utilising ICT in construction projects are available from IT Construction Best Practice (ITCBP) and IT Construction Forum publications. For some brief guidance on project management software see Briefing Note 3.04.

Project planning

The project master development schedule should be developed and agreed with the client and the project team, such that it represents the information available at the time for the whole project lifecycle. The master development schedule should include suitable contingencies to reasonably accommodate future developments in the detailed schedules for each of the subsequent stages. These developments typically include reasonably foreseeable scope changes and/or delays to activities as well as an allowance for design/scope evolution.

Increasing levels of detail should be added in a controlled manner – usually before each new stage of the project and as the necessary parameters are established (see Briefing Note 3.06). Care should be taken when administering increasing detail, such that the addition of detail is not confused with changes. Similarly, where increasing detail is added to subsequent phases, the overall period should either fit within the high-level period identified in the master development schedule, or where this is not possible, the master development schedule will require change and hence suitable administration.

As the project develops, activities which are on the critical path may change; During the early development stage, the critical activities are likely to be ones such as applying for and obtaining statutory approvals, external consultations and enquiries; legal and funding negotiations and any other third-party agreements. It is useful to understand which activities are critical in each phase and to allow a suitable contingency for them.

It is the project manager's responsibility to coordinate the projects' schedules – where more detailed schedules are developed by process owners or specialists; it is similarly the project manager's responsibility to make sure that they are suitably coordinated with the master development schedule. Key coordination activities include making sure that the schedules are developed and issued in a coherent and timely fashion; identification of suitable contingencies to accommodate reasonably foreseeable risks; to monitor, respond to and report on progress as well as to initiate necessary action to rectify potential or actual non-compliance.

Cost planning and controls

A development budget study is undertaken to determine the total costs and returns expected from the project. A cost plan is prepared to include all construction costs and all other items of project cost including professional fees and contingency. All costs included in the cost plan will also be included in the development budget in addition to the developer's returns and other extraneous items such as project insurance, surveys and agent's fees or other specialist advisors.

The objective of the cost plan is to allocate the budget to the main elements of the project to provide a basis for cost control. The terms *budget* and *cost plan* are often regarded as synonymous. However, the difference is that the *budget* is the limit of expenditure defined for the project, whereas the *cost plan* is the definition of what the money will be spent on and when. The cost plan should, therefore, include the best possible estimate of the cash flow for the project and should also set targets for the future running costs of the facility. The cost plan should cover all stages of the project and will be the essential reference against which the project costs are managed.

The method used to determine the budget will vary at different stages of the project, although the degree of certainty should increase as more project elements become better defined. The budget should be based on the client's business case and should change only if the business case changes. The aim of cost control is to produce the best possible building within the budget.

The cost plan provides the basis for a cash flow plan, based on the master development schedule, allocating expenditure and income to each period of the client's financial year. The expenditures should be given at a stated base-date level and at out-turn levels based on a stated forecast of inflation. A cash flow histogram and cumulative expenditure graph are shown in Figure 3.4.

Operational cost targets should be established for the various categories of running costs associated with the facility. This should accompany the capital cost plan and

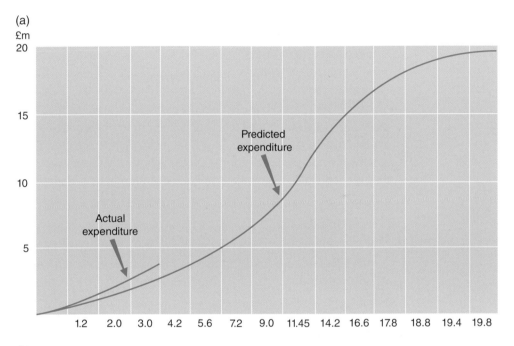

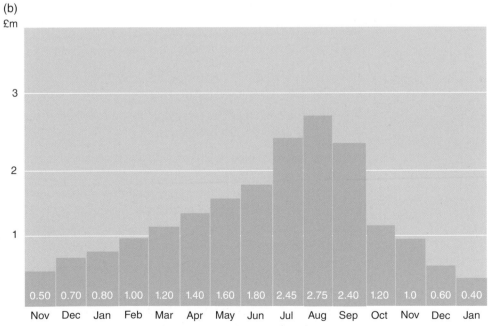

Figure 3.4 Examples of (a) construction expenditure graph and (b) cash flow histogram.

be included in the brief to consultants. The importance of revenue, grants and tax planning for capital allowances must also be taken into consideration.

When the cost plan is in place, it serves as the reference point for the monitoring and control of costs throughout a project. The list which follows should be used as an aid in setting up detailed cost-control procedures for all stages of a project.

Cost control

The objective of cost control is to manage the delivery of the project within the approved budget. Regular cost reporting will facilitate, at all times, the best possible estimate of:

- established project cost to date
- anticipated final cost of the project
- future cash flow

In addition, cost reporting may include assessments of:

- ongoing risks to costs
- costs in the use of the completed facility
- potential savings

Monitoring of expenditure to any particular date does not exert any control over future expenditure and, therefore, the final cost of the project. Effective cost control is achieved when the whole of the project team has the correct attitude to cost control, that is, one which will enable fulfilment of the client's objectives.

Effective cost control will require the following actions to be taken:

- Establishing that all decisions taken during design and construction are based on a forecast of the cost implications of the alternatives being considered, and that no decisions are taken whose cost implications would cause the total budget to be exceeded.

- Encouraging the project team to design within the cost plan, at all stages, and adopt the variation/change and design development control procedures for the project. It is generally acknowledged that 80% of cost is determined by design and 20% by construction. It is important that the project team is aware that no member of the team has the authority to increase costs on its section or element of the work. Savings on another must always balance increased costs on one item.

- Regularly updating and reissuing the cost plan and variation orders causing any alterations to the brief.

- Adjusting the cash flow plan resulting from alterations in the target cost, the master schedule or the forecast of inflation.

- Updating the cost plan in liaison with the project team as design and construction progresses. At all times it should comprise the best possible estimate of the final cost of the project and of the future cash flow. Adherence to design freezes will aid cost control. (Updating also means adding detail as more information about the work is assembled, replacing cost forecasts with more accurate ones or actual costs whenever better information can be obtained.)

- As part of risk management, reviewing contingency and risk allowances at intervals and reporting the assessments is essential. Development of the cost plan should not involve increasing the total cost.

- Checking that the agreed change management process is strictly followed at all stages of the project is very important (see Briefing Note 3.13).

- Arranging that the contractor is given the correct information at the correct time in order to minimise claims. Any anticipated or expected claims should be reported to the client and included in the regular cost reports.

- Contingency money based on a thorough evaluation of the risks is available to pay for events which are unforeseen and unforeseeable. It should not be used to cover changes in the specification or in the client's requirements or for variations resulting from errors or omissions. Should the consultants consider that there is no alternative but to exceed the budget, a written request to the client must be submitted and correct authorisation received. This must include the following:

  - details of variations leading to the request

  - confirmation that the variations are essential

  - confirmation that compensating savings are not possible without having an unacceptable effect on the quality or function of the completed project

- Submitting regular, up-to-date and accurate cost reports to keep the client well informed of the current budgetary and cost situation.

- Establishing that all parties are clear about the meaning of each entry in the cost report. No data should be incorrectly entered into the budget report or any incorrect deductions made from them.

- Ensuring that the project costs are always reported against the original approved budget. Any subsequent variations to the budget must be clearly indicated in the cost reports.

- Plotting actual expenditure against predicted to give an indication of the project's progress (see Figure 3.4).

Design management process (managing the design delivery)

The design management process includes the management of all project related design activities, people, processes, and resources such as:

- enabling the effective flow and production of design information

- contributing to achieving the successful delivery of the completed project, on time, on budget and in fulfilment of the client's requirements on quality and function in a sustainable manner

- delivering value through integration, planning, coordination, reduction of risk and innovation

- achieving through collaborative and integrated working and value management processes

However, there is much debate over the scope and responsibility for managing the design. This is because while the design team and the DTL (design team leader) are usually ultimately responsible for the production and drawing of the design, ensuring that the drawings are delivered to the project team and the relevant supply chain is usually the responsibility of the DDM (design delivery manager). This process of design delivery is an activity of project management.

Where the project is carried out under a design and build contract, the DDM may be part of the contractor`s organisation or the associated design company.

There are many publications on design management and to add yet another is not the intention of this document; there are two, however, which are strongly recommended:

3 Strategy

1. The RICS publication 'Managing the design delivery' 1st edition, guidance note (GN 76/2012)

2. The CIOB publication 'The design manager's handbook'.

Further guidance has also been provided in Briefing Note 3.17 – design management process.

Risk identification and management

Compared with many other industries, the construction projects are likely to be subject to more risks due to the unique features of activities, such as long duration, complicated processes, changeable environment, financial intensity and dynamic organisational features. Hence, effective risk identification and management is paramount for success of any construction project. Along with identification of risks, care should be taken to identify the most appropriate owner of the risk. Transferring all risks over to the contractor in most circumstances is unlikely to generate the best value solution and does very little to foster a collaborative environment. The risk identification and management process should have the best interest of the project as its core objective and it is the responsibility of the project manager to ensure conformance to this objective. Briefing Note 3.03 provides further guidance on project risk assessment.

Environmental management and controls

Environmental statements

Environmental concerns will increasingly affect our projects. This is especially the case with the pressure to develop brownfield sites and reuse old sites (e.g. the guidance as set out in NPPF (2012)). The cost of addressing contaminants or other environmental issues can add significant costs and increase the duration of project. Planning authorities are also more likely to instruct environment studies and restraints as part of the planning process, all of which must be incorporated into the project during the construction stage. It is the project manager that has overall responsibility to ensure compliance with these aims, objectives and constraints. The project manager will need to:

*   understand and act on the environmental impact assessment; see Briefing Note 2.03.

*   ensure proper environmental advice is available

*   ensure that the contractor is complying with the environmental statement; see Briefing Note 2.03

*   seek and ensure action by the contractor of any remedial actions should they be necessary to comply with environmental considerations

Contractor's environmental management systems

The contractor must establish his own environmental management systems (EMS), but it is for the project manager to ensure that it is being managed properly and is progressing sufficiently to achieve all EMS objectives. Therefore, the project manager should:

*   receive details of the contractor's EMS and the environmental plan (EP) specific to the project

*   ensure that the contractor has set up all necessary procedures and structure to manage the EMS and implement the objectives of the EP

*   check that the contractor's environment management plan matches the aims and objectives of the environmental statement

- agree with the contractor any further aims, specific targets or initiatives that will maximise sustainability of the project and minimise the detrimental impact of the construction process

- proactively monitor the progress of the contractor to maintain his proposals and objectives

Stakeholder management

Traditionally, the main participants in a construction project coalition are the client, the designers and the contractor with the project manager leading the coalition. The interactions and inter-relationships between these participants largely determine the overall performance of a construction project and have the crucial responsibility for delivering a project to successful completion. However, looking upstream and downstream in the construction project lifecycle, there are multiple attributes that contribute to the success of a project, and these are influenced by a variety of decisions made by various individuals, bodies and organisations. These internal and external participants are recognised as stakeholders who are actively involved in the project or whose interests may be positively or negatively affected as a result of project execution.

The feasibility stage discussed the process to identify the relevant stakeholders as appropriate in the project context and outlined the tools that can be used to undertake stakeholders mapping and management.

Internal stakeholders are people who have legal contact with the client and those clustered around the client on the demand side (employees, customers, end users and financiers) and on the supply side (architect, engineers, contractors, trade contractors and material suppliers). The external stakeholders comprised private and public sector representatives. The private entities can potentially be from the local residents, landowners, environmentalists and local pressure groups, whereas the public entities can be from regulatory agencies, and local and national government.

It is the responsibility of the project manager to ensure that an appropriate stakeholder mapping is undertaken, with the relevant processes necessary for management of each and every stakeholder depending on their position on the power/interest matrix, perhaps utilising a RACI (Report, Advise, Consult, Inform) or similar model.

Quality management

It is the project manager's role to set up and implement an appropriate process to manage project quality. From the quality policy defined in the project brief, the development of a quality strategy should lead to a quality plan setting out the parameters for the designers and for the appointment of contractors. Quality control then becomes the responsibility of the contractor, subcontractors and suppliers operating within the agreed quality plan. The plan itself should establish the type and extent of independent quality auditing (particularly for off-site production of components) and the timing of inspections and procedures for 'signing off' completed work.

It is the responsibility of the design team and other relevant consultants to specify the goods, materials and services to be incorporated in the project, using the relevant British Standards, codes of practice and '**Board of Agrément**' criteria or other appropriate standards.

The achievement of these standards rests with the appointed main contractor. When interviewing contractors at the pre-tender stage, the project manager will seek confirmation that each company has a positive and proactive policy towards the control of quality, a policy which will be reflected in all of its operations off or on the site.

3 Strategy

Commissioning strategy

Typically commissioning information, guidance and resources are linked with three broad principles:

* *determine project performance requirements*

* *plan the commissioning process*

* *document compliance and acceptance*

It is important to note that all three principles are applied over the lifespan of a capital design and construction project, and that it takes a multidisciplined effort involving clients, design professionals, construction managers and commissioning providers to achieve optimal results from the commissioning process.

It is important to start the commissioning process early and to bring the commissioning agent or entity on board during or before schematic design. This early involvement is critical for the timely and useful development of the Owner's Project Requirements (OPR), the subsequent design team Basis of Design (BOD) and the beginning of the Operations & Maintenance (O&M)Systems Manual. If these tasks are left until later in the process and 'reverse engineered' to match the design, their usefulness as catalysts for dialogue and quality tracking tools would be compromised.

Appointing the commissioning agent immediately after the designers allows the commissioning agent to become familiar with existing programming documents and when the Services Manual is started at this early stage, the inclusion of O&M requirements is ensured. The inclusion of O&M in the early stage project programming is the key to the long-term persistence of the energy efficiency and equipment longevity strategies built into the design.

Selection and appointment of project team consultants

The project manager in consultation with the client will decide on and implement a selection procedure for members of the protect team, together with contractors and other consultants. These may then be appointed on behalf of the client. The extent of contractors and consultants will be determined, to some extent, by the procurement route selected. In connection with the project team, two common arrangements for appointment are:

* separate appointment of independent service providers

* single appointment of a team of service providers or a lead organisation for the provision of all services

It is important that the members of the project team should be as compatible as possible, both in temperament and in working methods if the project is to have the greatest likelihood of success. Therefore, selection should be based on balancing quality, compatibility, schedule and price. The project team, consultants and contractors (who may or may not also be suppliers within the supply chain) may be appointed through a process of short-listing and structured interviews or through a competitive tendering procedure. This may be through European Union (EU) procurement procedures, which may be mandatory for publicly funded projects depending on the size of the project (see Briefing Note 3.11 for a brief guidance to EU procurement rules). The project manager needs to be fully informed on all issues related to the procurement process and advise the client accordingly.

Table 3.2 Appointment of the project team consultants

Activity	Considerations
Selection and appointment of project manager	May be appointed at inception/feasibility stage
Agree criteria for team selection	Type of expertise and scope Budget fee Contractual procurement strategy
Define in detail each function	Extent of services required Coordinate with other professional agreements
Define roles and duties	Scope of work Roles and duties
Agree terms and conditions of engagement	Clients of standard conditions of engagement Programme, professional indemnity (PI) insurance, warranties
Selection of those invited to bid	Utilise relevant databases Agree list of consultants Decide selection criteria Format proposal required Agree content and fee
Carry out the selection process	Agree interview team Agree selection criteria Arrange interviews Utilise scoring system
Negotiate conditions of appointment	Provide client with analysis and selection recommendations Negotiate final conditions
Advise on final appointment	Issue letter of appointment Issue letters of rejection
Oversee formalities relating to PI insurance, warranties and building	Legal department to formalise Finance department for fees Legal documents issued

The client should be consulted on the formulation of shortlists and should be invited to attend any interviews. The process is set out in Table 3.2.

For the 'short-list' method, the project manager should formulate the short lists, convene and chair the interviews, record and assess the results and present a report and recommendation to the client for final decision (see Briefing Notes 3.09 and 3.10 for further guidance on the selection process).

Most professional firms are members of organisations that publish standard terms of appointment and codes of conduct. It is usual to appoint the project team members on the standard terms which are designed to provide a proper balance of risk and responsibility between the parties. Standard terms are capable of amendment by agreement (but any amendment should be considered very carefully before changes are made to avoid inconsistencies and conflicting terms). The project manager should advise against terms which impose uninsurable risks or unquantifiable costs on the consultants or are in conflict with their professional responsibilities or codes of conduct.

The project manager will issue to the appointed project team with the project handbook, the outline project brief and the master schedule together with the budget or cost plan. It is advisable that these elements are referred to in as much detail as is available at the time of the project team's appointment.

3 Strategy

Collaborative arrangements

Collaborative and partnering arrangement is a set of actions by project team by which risks can be distributed and conflict can be minimised. The intention is to provide 'win-win' outcomes for everyone involved. It is put into practice by having regular partnering workshops of all the key members of the project team to establish and foster cooperative ways of working aimed at improving performance. In broad terms, partnering teams agree mutual objectives that take account of the interests of all the parties; establish cooperative methods of decision-making including procedures for resolving problems quickly; and identify actions to achieve specific improvements to normal performance. The workshops take place throughout the project usually under the guidance of an independent partnering facilitator.

There is considerable evidence to show that partnering can bring significant net benefits in the form of reduced costs, improved quality and shortened timetables. However, these benefits involve additional costs in ensuring the selected project team members are willing to work cooperatively and the expense of running partnering workshops. The additional costs arise during the early stages of projects while the benefits, which depend on partnering being successful, accrue during the later stages. The likely costs and benefits of partnering should be considered by all clients. It is very likely to provide benefits for clients with phased projects or who undertake programs of similar projects, but it has been shown to provide benefits on a one-off project. Partnering of the supply chains that produce key elements should also be considered on all except the smallest and simplest projects (see Briefing Note 3.14).

Framework arrangements

Rather than uniquely tender goods or services on individual projects, some large clients prefer to establish framework agreements with preferred suppliers. Frameworks can be described as agreements to provide both goods and services on predefined and specified terms and conditions with a selected number of suppliers (e.g. consultants, designers and contractors). Frameworks range from a mechanism to assemble a limited number of prequalified suppliers who will then be asked to bid for projects to arrangements where specific teams of suppliers are guaranteed a regular workload in return for continuous improvement in delivery of their product. There are a wide variety of framework procurement contracts available for use. Brief guidance on framework contracts has been provided in Briefing Note 3.08.

Private public partnership/private finance initiative (PPP/PFI)

A PPP project is any alliance between public bodies, local authorities or central government, and private companies to deliver a project. A PFI is one type of PPP. A PFI is a more specific and formal long-term partnership, covering both the capital asset and the services that form a project. Typically, under a PFI regime, the public sector sets a level of service and the private sector operator provides the services in return for a charge, so the public sector body is able to finance the project over the term of a contract, often 25 or 30 years. Funding, in the form of PFI credits, may be available from central government to support the capital element. Though there is a significant variance on a large number of factors, many PFI/PPP projects have the following features in common:

- long project life, usually at least five years, sometimes many more

- a substantial initial capital spend which is recovered over the project's life

- financed with high levels of borrowed capital

- income and risk-sharing determined by a long-term contract

Other forms of PPP projects may include arrangements such as DBFO (design built finance operate), DBO (design built operate), Concessions, JV (joint ventures), outsourcing and similar.

Further information relating to PPP projects has been incorporated in Briefing Note 3.15.

Procurement strategy

In the context of this Code of Practice, procurement should be considered to be the process of identification, selection and commissioning of the contributions required for the construction phase of the project. The alternative methods of procurement referred to reflect the different organisational and contractual arrangements which can be made to ensure that the appropriate contributions are properly commissioned and that the interests of the client are safeguarded.

The various procurement options available reflect fundamental differences in the allocation of risk and responsibility to match the characteristics of different projects; therefore selection of the procurement option must be given strategic consideration. The project manager should advise on the relative benefits and disadvantages of each option, related to the particular circumstances of the project, for the benefit of the client.

The final choice of procurement method should be made on the basis of the characteristics of the project, the client and his requirements. The selection of method should be made when consideration is being given to the appointment of design and other specialist consultants because each option can have a different impact on the terms of appointment of the members of the project team.

The various procurement methods which may be pursued can be broadly classified under four headings:

- traditional
- design and build
- management contracting
- construction management

Each method has its own variations. No method is best in all circumstances. They bring different degrees of certainty and risk towards construction and development.

It is important to note that works contract selection will be influenced by the procurement method chosen and must be harmonious with the procurement strategy.

Traditional

The contractor builds a client-designed scope of work within a time period for a lump sum. The client remains responsible for the design and the performance of his consultants under the building contract. The client appoints a design team, including a quantity surveyor responsible for financial and contractual advice. A building contractor is appointed, usually after a tender process, and typically based on one of the standard forms of contract, to carry out the construction. The tender process can be based on complete design information or partial design information plus provisional guidance if an early construction start is required.

Some traditional contracts also allow for design responsibility for certain elements of the works to be passed to the main contractor, sometimes referred to as contractor's design portions (CDPs). This is particularly relevant where there are specialist elements of the works that need to be designed and installed by a specialist subcontractor as part of the main contractors' works.

Design and build

The client appoints a building contractor, typically on a standard form of contract to carry out at least some of the design and to complete construction within a time period for a lump sum. The contractor is responsible for the design elements and construction as defined in formal documentation known as the client's requirements, which usually impose the same design responsibility as would be imposed on a designing consultant, that of 'due skill and care'; it does not normally require the contractor to warrant 'fitness for purpose'. The appointment may be made after a tendering process incorporating variations on the method or through negotiation. The client may appoint consultants to prepare the client's requirements, which may involve varying degrees of design and after-contract to oversee matters on his behalf. Difficulties can arise in distinguishing changes in the client's requirements (for which the client is liable) from design development (for which the contractor is liable).

If the design-and-build contractor is appointed after completion of a part of the design, the appointments of the design team may be formally passed on (contractually novated) to the design-and-build contractor. The purpose of the novation is to secure continuity and consistency of quality through to the production information stage when working under the direction of the contractor. However, in practice, research has shown the potential for conflicts and poor quality often remains.

Prime contracting is an extension of the design-and-build concept. The prime contractor will be expected to have a well-established relationship with a supply chain of reliable suppliers. The prime contractor coordinates and project manages throughout the design and construction period to provide a facility which is fit for the specified purpose, and meets its predicted through-life costs. The prime contractor is paid all actual costs plus profit incurred in respect of measured work and design fees; it is only at risk in respect of its staff and preliminaries.

Management contracting

The client appoints a design team with responsibilities, as in the traditional method, and augmented by a management contractor whose expertise and advice is available throughout the design development and procurement processes. Specialist works subcontractors, who are contracted to the management contractor on terms approved by the contract administrator, carry out the construction. The appointments of the management contractor and the trade subcontractors are usually made on standard contract forms. The management contractor is reimbursed all his costs and paid a percentage on project costs in the form of a guaranteed profit or fee.

Construction management

Construction management requires that the specialist works contractors are contracted to the client directly, involving the construction manager as a member of the project team acting as an agent and not a principal, to concentrate on the organisation and management of the construction operations. The project team, including the construction manager, is responsible for all financial administration associated with the works. The construction manager is paid an agreed fee to cover the costs of staff and overheads. This is generally considered to be the least adversarial form of contract and is often invoked when design needs to run in parallel with construction. This method is at its best with a hands-on responsive client who can make decisions quickly.

Hybrid procurement approaches

Due to the varied nature of project requirements and constraints, it is often not practical to adhere to the forms of procurements as identified above which results in clients opting for more of a 'mix and match' route which results in utilising a hybrid procurement option. The choice of the hybrid ways, as highlighted later, will depend on:

- client requirements

- client experience

- development and availability of project information

- delivery time scales

Two-stage tender: This is a variation of the traditional (design – bid – build) approach where the contractor's tenders are based on a partially developed consultant's design (stage 1 tender). The contractor then assists with the final development of the design and tender documents, against which tenders for the construction works are prepared (stage 2 tender). Whoever put forward the first stage tender has the opportunity to tender or negotiate the second (construction) stage. This approach increases the risks of an increase in overall price and a less certain completion date but contractor involvement is likely to increase the likelihood that both these criteria are realistically established.

Develop and construct: This is a variant to the design-and-build approach, where the client has the design prepared to concept or scheme design stage and the contractor takes on 'finishing off' the design and construction. The contractor may re-employ (through novation) the original designers to complete the design.

Package deal: This is a variant to the design-and-build approach, where the contractor provides an off-the-shelf building. The building type is often modular so that its size can be adjusted. Typical examples are farm, factory, warehouse, office buildings and school class rooms.

There are also a number of other possible combinations of hybrid procurement options as well.

Innovative form of procurement

Based on the findings of the Government Construction Task Group Report (2012),[6] the Government Construction Strategy set out a clear commitment to trialing new models of procurement that include principles of early supplier engagement, transparency of cost, integrated team working and collaborative working. A draft guidance for these models had been published in early 2014 which aimed to set out three key procurement routes:

Two Stage Open Book: Using Two Stage Open Book a client invites prospective integrated teams to bid for a project based on their ability to deliver an outline brief and cost benchmark. Following the first stage competition, the appointed team works alongside the client to build up a proposal, the construction contract being awarded at the second stage. This differs from Cost Led Procurement in reducing industry bidding costs, enabling faster mobilisation and in providing the opportunity for clients to work earlier with a single integrated team.

Cost Led Procurement: Applying the Cost Led Procurement process, a client can use their knowledge of costs to set a challenging cost ceiling and output specification against which the supply chain can bring experience and innovation to bear in a competitive framework environment. On frameworks with a series of similar capital

---

[6] See https://www.gov.uk/government/collections/new-models-of-construction-procurement (accessed May 2014).

projects, CLP provides the opportunity to continually improve on the unit costs of the programme working collaboratively with the supply chain.

Integrated Project Insurance: The Integrated Project Insurance (IPI) model offers clients the opportunity to create a holistic and integrated project team (an 'Alliance Board') to eliminate the "blame/claim" culture. The innovative "integrated project insurance" package limits the risk for the individual members of the team, fosters joint ownership of the project, and thereby reduces the likelihood of overrunning in terms of cost and time.

Characteristics of procurement options

The characteristics of four basic forms of procurement are explained in detail in Briefing Note 3.07.

Procuring the supply chain

When selecting the key members of the supply chain, not only should the key contractors be appointed as soon as is feasible, depending on the procurement route selected, but key suppliers, especially those who may have a design element within their requirement, should likewise be appointed at an early stage of the project. This should enable the following to be achieved:

- resolution of the buildability issues at design stage

- choice of the most efficient materials to be used

- advice to the client as to costing (practical, rather than rates)

- coordination of the 'specialist designers' with main design team

- understanding of the client's needs in all areas, which will assist in the construction process leading to the quality and required product for the client

- highlighting and incorporation of relevant health and safety issues which must be taken into account in the design including obligations under CDM regulations

Responsible sourcing

Responsible sourcing relates to the management of sustainability issues associated with materials in the construction supply chain, often from an ethical perspective. During the procurement process, in addition to the obligations under CDM 2007 regulations, efforts must be made to ensure that the supply chain, at as many levels as practicable, can demonstrate through processes and procedures, a safe, sustainable and ethical sourcing philosophy.

Tender procedure

The master development schedule will indicate the time allowed for procurement and design, will show activities such as tender interviews, tendering and selection; scope, release dates, approval periods, cost checking and appropriate documentation, etc. These activities may include the following (see also Figure 3.5):

- Checking that the various documents are produced at the appropriate times, including those for pre-main construction works (e.g. demolition, site clearance, access and hoarding) and ensuring that they contain any special terms required by the client and local statutory authorities for activities such as archaeology and environmental

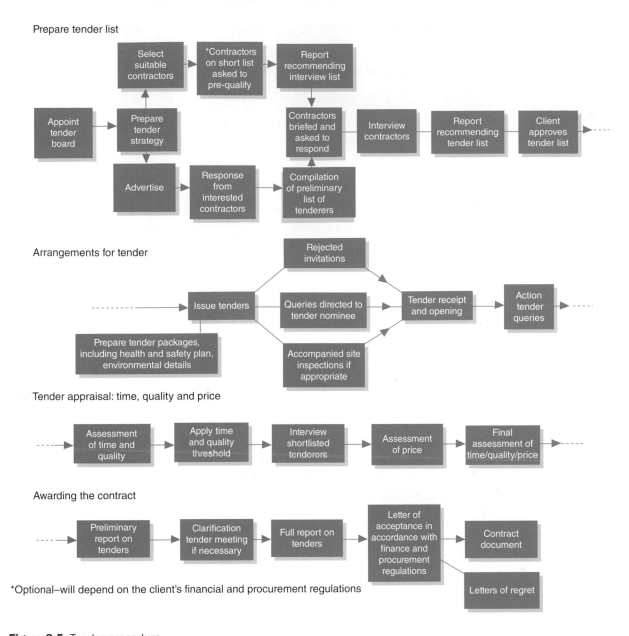

Prepare tender list

Arrangements for tender

Tender appraisal: time, quality and price

Awarding the contract

*Optional—will depend on the client's financial and procurement regulations

Figure 3.5 Tender procedure.

investigations. In conjunction with the relevant consultants, preparing lists of firms to be invited to tender for the elements of the work (pre-qualifying process). Obtaining confirmation that the listed firms will be prepared to submit their offers at specified dates, taking up references and/or interviewing prospective supply chain companies, including relevant consultants.

• Ensuring that all appropriate health and safety/CDM references are made in the documentation together with any relevant statutory authority references.

• Checking, in liaison with other project team members, that all contract documentation is coordinated, especially where aspects of design are involved and confirming that, where appropriate, warranties are secured.

• Receiving reports on supply chain procurement, together with method statements.

• Interviewing successful supply chain applicants, if necessary, to clarify any special conditions and to meet significant leading personnel.

3 Strategy

- Arranging for formal acceptance of successful supply chain applicants as appropriate and issuing award letters.

- Selection should be based on balancing quality, schedule and price.

- Initiating action if price submissions are outside of the budget.

- Ensuring that the client understands the nature and terms of the construction, particularly those in relation to possession and payment terms, and that possession of the site can be given to the contractor on the date set out in the tender.

- Arranging for formal signing and exchange of contracts.

In view of the EU directives on procurement, negotiated tendering is also being undertaken as an option to secure best value for money. See Briefing Note 3.11 for further information about EU procurement rules.

Procurement under EU directives

EU Directives (Directive 2004/17/EC and 2004/18/EC dated 31 March 2004) enacted by UK (The Utilities Contracts Regulations 2006 and The Public Contracts Regulations 2006 enforced 31 January 2006) require that for public procurement above a certain monetary threshold, contracts must be advertised in the *Official Journal of the European Union* (OJEU) and there are other detailed rules that must be followed. This is irrespective of the procurement route chosen. Further advice and guidance on this has been provided in Briefing Note 3.11.

e-Procurement

Electronic procurement (e-procurement) is described to be the use of electronic tools and systems to increase efficiency and reduce costs during procurement. The EU Consolidated Directives and EU Invoicing Directives provide clear directions for enabling e-procurement, including encouraging online-only processes whereby tenderers can compete for contracts. In addition to key public sector clients such as the Ministry of Defence, the National Health Service and local authorities, private sector clients are also increasingly utilising this process of procurement. Further guidance on e-procurement has been provided in Briefing Note 3.16.

Employer's requirement document

The purpose of the employer's requirements document is to set out precisely the requirements of the client to facilitate contractors to price the works accurately. While at the same time removing any risk of unforeseen costs for elements of work that have not been accounted. The size of the project and the complexity of the business requirements will be the driving factor as to what the employer's requirements should contain.

This document is typically utilised in design–and-build form of procurement.

The employer's requirements document is developed for individual projects, but would typically incorporate the following pre contract information. Once completed these documents would form the basis of the contract between the employer and the contractor.

Employer's requirements:

- project overview

- pre-acquisition or survey reports forming part of the building due diligence

- building appraisal report

- proposed form of building contract and completed appendices

- contract specific terms & conditions

- proposed tender and construction schedules

- detailed scope of works

- requirements for the works to comply with statutory obligations

- health & safety compliance

- acoustic requirements

- mechanical & electrical performance specification

- voice & data requirements

- audiovisual specification

- wall and floor finishes schedules

- furniture specification and space standards

- proposed drawing package

- concept and detailed design proposals

- form of tender

Facility management strategy/considerations

From a whole lifecycle perspective as well as from an operational perspective, it is essential that adequate considerations are given to the proposed facilities management (FM) requirements (in context of resource availability vs. resource requirement) prior to any significant design development is undertaken. It is likely that key design decisions will be formulated upon the future FM requirements and resources availability. It is the responsibility of the project manager to ensure that appropriate stakeholders, particularly those who will be responsible for satisfying the future FM needs of the delivery item, are consulted and taken on board prior to embarking upon the design development stage.

Project execution plan

The project execution plan (PEP) is the core document for the management of a project. It is a statement of policies and procedures defined by the project sponsor; although it is usually developed by the project manager for the project sponsor's approval. It sets out in a structured format the project scope, objectives and relative priorities.

This is a live document that enforces discipline and planning having a wider circulation than the project design team. It forms a basis for:

- sign off by the client body at the end of the feasibility and strategy phases

- an aid for funding

- a *modus operandum* for the project team

- an information and 'catch up' document for prospective contractors

Some of the confidential information in the client version will be taken out of the published version to other parties.

Checklist for the PEP

Does the PEP:

- Include plans, procedures and control processes for project implementation and for monitoring and reporting progress?

- Define the role and responsibilities of all project participants, and is a means of ensuring that everyone understands, accepts and carries out their responsibilities?

- Set out the mechanisms for quality control, audit, review and feedback, by defining the reporting and meeting requirements, and, where appropriate, the criteria for independent external review?

Essential contents

A large part of PEP may be standardised, but the standard will need to be modified to meet the particular circumstances of each project. A typical PEP might cover the items listed next, although some may appear under a number of headings with a cross-reference system employed to avoid duplication:

- project definition and brief

- statement of objective

- the business plan with costs, revenues and cash flow projections including borrowings interest and tax calculations

- market predictions and assumptions in respect of revenue and returns

- functional and aesthetic brief

- client management and limits of authority including the project manager

- financial procedures and delegated authority to place orders

- development strategy and procurement route

- statutory approvals

- risk assessment

- project scheduling and phasing

- the scope content of each consultant appointment

- reconciled concept design and budget

- method statement for design development, package design and tendering, construction, commissioning and handover, and operation

- safety and environmental issues, such as the construction design and management regulations, carbon emissions and energy targets

- management of information systems including document management systems.

- quality assurance

- Post-project evaluation

The PEP will change as a project progresses through its design and construction stages. It should be a dynamic document regularly updated and referred to as a communication tool as well as a control reference.

Approval to PEP

Typically it is the responsibility of the project manager to ensure that the key project stakeholders establish the PEP approval process which will be linked with the project milestones. The project manager must ensure that the PEP is approved progressively as the document is developed through the project life cycle in accordance with the agreed procedures.

BIM strategy

The objectives for this stage are to ensure that the project is set up correctly to ensure success.

In terms of BIM, a clear plan needs to be agreed with all project stakeholders to ensure that maximum benefits and efficiencies will be obtained for the client, the project and the team.

The plan at this stage should include:

- Clear statement and understanding of the objectives for the use of BIM on the project, for example why is BIM being used? What are the desired outcomes?

- Development of the project BIM execution plan, which will include assessment of capabilities of the team and protocols for integrated working, including software, data storage and communications. Members of the supply chain may initially require additional support to work in a BIM environment, and this should be factored in at the earliest opportunity.

- Outputs required by the client –for example, is an FM related BIM data set required at handover? Are interim outputs required, for example, COBie? For what purpose?

- Matrix of responsibilities – including ownership of model elements and data

- Appointment of the BIM manager or coordinator, with clear responsibilities. This could be one of the existing team or an external consultant depending on capability and experience

- As part of the BIM execution plan this will set down the model development process in terms of level of detail progression, authorised uses of the model data at different stages and anticipated outputs, for example, COBie.

- Process for design development, including which BIM tools are to be used (e.g. clash detection, 4D, 5D, environmental analysis. etc.) and the iterative processes required for compliance, including management of interactive design team meetings.

As BIM usage on the project develops, the execution plan and processes require regular review to ensure that BIM processes are working as efficiently as possible.

If the requirements and usage of BIM on the project are understood and clearly communicated at the beginning of the project. then it is relatively straightforward to develop the project BIM to achieve the desired outcomes. Conversely, if this has not been achieved in the early stages, it is probable that the BIM process will not be effective as it could be.

Further information and resources have been outlined in Briefing Note 3.05.

Briefing Note 3.01   Health and safety in construction including CDM guidance

Generally, the laws governing health and safety relate to all construction activities (including design) and are not industry specific. There are several Acts and Regulations involved and information on how to access these is provided at the end of this appendix.

Some of the principal Acts which deal with health, safety and welfare in construction are as follows:

- Health and safety at Work etc. Act 1974

- Mines and Quarries Act 1954

- Factories Act 1961

- Offices, Shops and Railways Premises Act 1963

- Employers Liability Acts – various

- Control of Pollution Act 1989

- Highway Act 1980

- New Roads and Streetworks Act 1991

- Corporate Manslaughter and Corporate Homicide Act 2007

The fundamental Act governing health and safety in construction is the Health and safety at Work etc. Act 1974. This act has some 62 separate Regulations and it is not possible to deal with such a large subject area here, however, the principal regulations of this Act, which affect design and construction, are:

- Management of Health and Safety at Work Regulations 1999 amended 2006

- Construction (Design and Management) Regulations 2007 (known as the CDM Regulations)

- The Work at Height Regulations 2005 amended 2007.

- Some other related regulations and guides are

- Site Waste Management Plans Regulations 2008

- Reporting of Injuries, Diseases and Dangerous Occurrences Regulations (RIDDOR) 1995

- The Control of Major Accident Hazards Regulations 1999 (COMAH) amended 2005

- The Chemicals (Hazard Information and Packaging for Supply) Regulations 2003 (CHIP 3)

- The Health and Safety (Display Screen Equipment) Regulations 1992

- COSHH (Control of Substances Hazardous to Health) Regulations 2002: Provision and Use of Work Equipment Regulations (PUWER 98)

- Lifting Operations and Lifting Equipment Regulations (LOLER 98)

- Personal Protective Equipment at Work Regulations 1992

- Signposts to the Health and Safety (Safety, Signs and Signals) Regulations 1996

- Control of Asbestos Regulations 2006

CDM 2007 Regulations

The Construction (Design and Management) Regulations 1994 is a Statutory Instrument that was introduced in compliance with the European Directives signifying a major change in the focus of the industry in terms of management of health and safety. For the first time, specific and explicit duties were placed on the client and the designers in context of management of health and safety. These regulations were updated in 2007.

The 2007 CDM Regulations have been introduced to provide a simplified regulatory structure with greater focus on planning and management of construction activities, with strengthened requirements on cooperation and coordination among all the key parties.

CDM 2007 contains five parts:

- Part 1 – introduction

- Part 2 – general management duties applying to all construction projects

- Part 3 – additional duties where projects are notifiable

- Part 4 – duties relating to health and safety on construction sites

- Part 5 – general

In addition, it is also supported by a CDM 2007 Approved Code of Practice (ACoP).

These regulations apply to all construction work, regardless size and duration. However, the 'notifiable' projects trigger additional duty holders and duties as identified in Part 3 of the regulations including:

- principal contractor

- CDM coordinator (the role of planning supervisor under CDM 1994 has been removed)

- notification to the Health and Safety Executive (HSE) (F10 form)

- construction phase plan

- health and safety file

Regardless of notification, most duties remain on clients, designers and contractors, and in most cases, there are additional duties.

Notifiable projects

The triggers for notification have also been simplified from CDM 1994. Under CDM 2007, notifiable construction projects are with a non-domestic client and involve either construction work lasting longer than 30 days or construction work involving 500 person-days.

Definition of client

CDM 2007 defines a client as an individual or an organisation who, in the course or furtherance of a business, has a construction project carried out by another or by themselves. This excludes domestic clients (i.e. someone who lives or will live in the premises where the work is carried out). The CDM client duties will still apply to domestic premises if the client is a local authority, landlord, housing association, charity, collective of leaseholders or any other trade or business.

For PFI or PPP projects, the project originators are the client at the start of the project until the special-purpose vehicle has been set up and has assumed the role of the client.

Role of clients under CDM 2007

The 2007 CDM Regulations do not, in the main, place new duties on the client, however:

- Existing duties under the old CDM regulations, as well as other relevant regulations, have been made explicit.

- Clarifications have been made with respect to how these duties should be exercised.

- Clients are accountable for the impact they have on health and safety.

- Clients are to ensure that a CDM coordinator is appointed to advise and coordinate activities on notifiable projects and adequate time and resources are provided to allow the project to be delivered safely.

- It is the responsibility of the client to provide key information to the designers and contractors; also it is for the client to arrange for any gaps in information to be filled in (e.g. commissioning an asbestos survey).

- Clients are to appoint a competent CDM coordinator and a competent principal contractor, and to ensure that the construction phase does not start unless the welfare facilities are in place and the construction phase health and safety plan is prepared.

- It is the responsibility of the client to retain and provide access to the health and safety file and revise it with any new information.

- For notifiable projects where no CDM coordinator or principal contractor is appointed, the client will be deemed to be the CDM coordinator and/or the principal contractor and be subject to their duties.

If the client makes a reasonable judgement that the contractor's management arrangements are suitable, taking account of the nature and risks of the project, and it is clearly based on evidence, clients will not be criticised if the arrangements subsequently prove to be inadequate or fail to be implemented without the client's knowledge.

Role of the CDM coordinator

CDM 2007 has created the new role of CDM coordinator which replaces the role of the planning supervisor under CDM 1994. The CDM coordinator is to advise the client on health and safety issues during design and planning phases of construction work and is only need to be appointed for notifiable projects. The key duties include:

- Advising the client about selecting competent designers and contractors – they do not have to approve the appointments.

- Helping to identify what information will be needed by designers and contractors.

- Coordinating the arrangements for health and safety of planning and design work – they do not have to check the designs, although they have to be satisfied that the hierarchy is addressed.

- Ensuring that the HSE is notified of the project (for notifiable projects only).

- Advising on the suitability of the initial construction phase plan – they do not have to approve or supervise the principal contractor's construction phase plan or work on site or approve risk assessments and method statements.

- Preparing a health and safety file.

The duties of a CDM coordinator can be carried out by a client, principal contractor, contractor, designer or a full-time CDM coordinator.

Role of designers

Under CDM 2007 a designer is someone who designs or specifies building work, including non-notifiable and domestic projects. This includes people who prepare drawings, design details, analysis and calculations, specifications and bills of quantities. In broader terms, these would include civil and structural engineers, building services engineers, material specifiers, temporary works designers, interior fit-out designers, clients who specify and design and build contractors. However, local authorities, etc. providing advice on relevant statutory requirements are not designers. If they require particular features that are not statutory then they are designers.

The key duties of the designers, under CDM 2007, are as follows:

- Ensure clients are aware of their duties

- Make sure they (the designer) are competent for the work they do

- Coordinate their work with others as necessary to manage risk

- Cooperate with CDM coordinator and others

- Provide information for the health and safety file

- Eliminate hazards from the construction, cleaning, maintenance, and proposed use (workplace only) and demolition of a structure

- Reduce risks from any remaining hazard

- Give collective risk reduction measures priority over individual measures

- Take account of the Workplace (Health, Safety & Welfare) Regulations 1992 when designing a workplace structure

- Provide information with the design to assist clients, other designers and contractors

- In particular, inform others of significant or unusual or 'not obvious' residual risks

- Risks which are not foreseeable do not need to be considered

- CDM 2007 does not require 'zero-risk' designs

- The amount of effort made to eliminate hazards should be proportionate to the risk

- Check that the client has appointed a CDM coordinator

- Only 'initial' design work is permitted until a CDM coordinator has been appointed: initial design can be considered to be no more than

- work within and beyond RIBA Stage C

- work within and beyond CIC Consultant Contract 2006 Stage 3 (draft as at August 2006)

- work beyond OGC Gateway 1

- work within and beyond ACE Agreement A (1) or B (1) 2002 Stage C3.

- Cooperate with the CDM coordinator, principal contractors and with other designers or contractors so all can comply with their CDM duties

- Provide relevant information for the health and safety file

Role of the principal contractor

The roles and responsibilities for the principal contractor have in fact been changed very little between CDM 1994 and CDM 2007. A principal contractor must be appointed as soon as practicable for all notifiable projects and the principal contractor should ensure that client is aware of appropriate duties, that the CDM coordinator has been appointed and that the HSE notified. Further responsibilities of the principal contractor include:

- Those they appoint, including all contractors and subcontractors, are competent.

- That the construction phase is properly planned, managed, monitored and resourced.

- That contractors are made aware of the minimum time allowed for planning and preparation.

- Providing relevant information to contractors.

- Ensuring safe working, coordination and cooperation between contractors.

- Ensuring that the construction phase health and safety plan is prepared and implemented. This plan needs to set out the organisation and arrangements for managing risk and coordinating work and should be tailored to the particular project and risks involved.

- Making sure that suitable welfare is available from the start of the construction phase.

- Ensuring that site rules as required are prepared and enforced.

- Providing reasonable direction to contractors including client-appointed contractors.

- Controlling access to the site to restrict unauthorised entry.

- Making the construction phase plan available to those who need it.

- Providing information promptly to the CDM coordinator for the health and safety file.

- Liaising with the CDM coordinator in relation to design and design changes.

- Ensuring all workers have been provided with suitable health and safety induction, information and training.

- Ensuring that the workforce is consulted about health and safety matters.

Under CDM 2007, a principal contractor does not have to

- provide training to workers they do not employ (it is the responsibility of individual contractors to train those they employ)

- undertake detailed supervision of contractors' work

Duties on contractors and self-employed workers

Check clients are aware of their duties:

- Plan, manage and monitor their own work to make sure that their workers are safe.

- Ensure that they and those they appoint are competent and adequately resourced.

- Inform any contractor that they engage, of the minimum amount of time they have for planning and preparation.

- Provide their workers (whether employed or self-employed) with any necessary information and training and induction.

- Report anything that they are aware of that is likely to endanger the health and safety of themselves or others.

- Ensure that any design work they do complies with CDM designer duties.

- Comply with the duties for site health and safety.

- Cooperate and coordinate with others working on the project.

- Consult the workforce.

- Not begin work unless they have taken reasonable steps to prevent unauthorised access to the site.

- Obtain specialist advice (e.g. from a structural engineer or occupational hygienist) where necessary.

- Check that a CDM coordinator has been appointed and that have been HSE notified before they start work (for notifiable projects only).

- Cooperate with the principal contractor, CDM coordinator and others working on the project.

- Inform the principal contractor about risks to others created by their work.

- Comply with any reasonable directions from the principal contractor.

- Work in accordance with the construction phase plan.

- Inform the principal contractor of the identity of any contractor they appoint or engage.

- Inform the principal contractor of any problems with the plan or risks identified during their work that have significant implications for the management of the project.

- Inform the principal contractor about any death, injury, condition or dangerous occurrence.

- Provide information for the health and safety file.

Duties to control site health and safety

Part 4 of CDM 2007 contains the duties to control specific construction health and safety risks. It has similar duties compared to the old Construction (Health, Safety and Welfare) Regulations 1996, which Part 4 replaces, in that:

- applies to all construction sites

- duties are on every contractor and every other person who controls construction work

- the wording and style has been updated and structure altered in parts, but retains most of the basic requirements of the original regulations

There are some changes, however:

- Good order now requires a site to be identified by suitable signs, be fenced off or both in accordance with the level of risk.

- New requirement to record in writing arrangements for demolition and dismantling.

- Excavation, cofferdam and caisson provisions have been extensively rewritten to make them more succinct and cohesive.

- Duties on reports and inspections have been restructured.

- Rest facilities, now requires seats with backs (specific requirement of the European directive – only required if replacing existing seating).

- Training and competence, specific requirements covered in the general part of the regulations.

- Requirements for doors and gates have been moved to Workplace (Health, Safety & Welfare) Regulations 1992.

- The provision to have equipment available to replace a rail vehicle on to its tracks has been removed.

- Prevention of drowning, provisions on a vessel's construction, maintenance and under control by competent person has been removed.

Competence and training

To be competent, an organisation or individual must have:

- Sufficient knowledge of the specific tasks to be undertaken and the risks which the work will entail.

- Sufficient experience and ability to carry out their duties in relation to the project; to recognise their limitations and take appropriate action in order to prevent harm to those carrying out construction work, or those affected by the work.

- All people who have duties under CDM 2007 should

  - take 'reasonable steps' to ensure persons who are appointed are competent

  - not arrange for, or instruct, a worker to carry out or manage design or construction work unless the worker is competent

  - not accept an appointment unless they are competent

Note that:

- The regulations apply to corporate and individual competence.

- Assessment should focus on the needs of the particular project and be proportionate to the risk, size and complexity of the work.

- CDM 2007 should streamline the competence assessment process.

- A key duty of the CDM coordinator is to advise the client about the competence of those employed by the client.

- Worker engagement is the participation by workers in decisions made by those in control of construction activities, in order that risk on site can be managed the most effective ways.

- Communication of the right information, to the right people at the right time to enable them to make appropriate decisions on health and safety issues relating to construction projects.

- The key CDM 2007 Regulations on worker engagement and consultation are:

  - Reg. 5: cooperation and Reg. 6 coordination

  - Reg. 10: client to provide information

  - Reg. 11: designers to provide information

- Reg. 13: contractors to provide information [including site induction, information on risks, site rules, imminent danger procedures, training as required by Reg. 13(2)(b) MHSWR].
- Provide employees with training when exposed to new/increased risks due to being transferred (new site), change of responsibilities, new equipment, new technology, new system of work, etc.
- For all projects, duty holders should:
  - provide information needed to carry out work without risk
  - provide site-specific induction
  - advise on findings from risk assessment
  - explain site rules
  - explain what to do if imminent danger
  - advise who is responsible for implementing health and safety on site.
- Allow for those who either cannot read or may not understand English.
- Workers have the duty to report anything that will endanger themselves or others.
- Worker safety representatives are entitled to employer-funded training.

Arrangements for serious danger under Reg. 8 MHSWR:

- communicate to workers what to do for a notifiable project the principal contractor must:
- make and maintain arrangements to ensure cooperation and consultation between themselves, contractors and workers
- to carry out consultation with workers.

CDM 2007: further advice

Further advice about the CDM 2007 Regulations is obtainable from:

- CDM 2007 Regulations and Approved Code of Practice (L144)
- HSE website: www.hse.gov.uk/construction/cdm.htm
- CDM 2007 industry guidance: www.cskills.org/supportbusiness/healthsafety/CDMRegs.

Design issues:

- www.dbp.org.uk
- www.dqi.org.uk
- www.cic.org.uk
- www.ciria.org.uk.

The Health & Safety Executive (HSE) have undertaken a consultation process which seeks views on HSE's proposal to replace the Construction (Design and Management) Regulations 2007 (CDM 2007) and withdraw the Approved Code of Practice. In line with an evaluation of CDM 2007 and a recommendation in Löfstedt's report "Reclaiming health and safety for all"[1]. The proposed revision is intended to aid clarity and reduce bureaucracy. The ACoP would be replaced with a suite of tailored guidance aimed at particular sectors, in particular smaller projects. Two thirds or more of fatal injuries in construction occur on small sites.

[1] Available at https://www.gov.uk/government/publications/reclaiming-health-and-safety-for-all-lofstedt-report-a-review-of-progress-one-year-on (accessed May 2014).

Briefing Note 3.02   Guidance on value management

Value management and value engineering

Value management (VM) and value engineering (VE) are techniques concerned with achieving 'value for money'. VE was pioneered by an American, Lawrence Miles, during the Second World War to gain maximum function (or utility) from limited resources. It is a systematic team–based approach to securing maximum value for money where

$$Value = \frac{Function}{Cost}$$

Thus, value can be increased by improved function or reduced cost. The technique involves identification of high-cost elements, determination of their function and critical examination of whether the function is needed and/or being achieved at lowest cost. In terms of projects, VE has the greatest influence and impact at the strategy/ design stage. It requires reliable and appropriate cost data and uses brainstorming workshops by a group of experts under the direction of a facilitator.

VM is similar to VE, but in terms of projects it focuses on the overall objectives and is most appropriate at the option identification and selection stage where the scope for maximising value is greatest.

Brainstorming forums involving those who would naturally contribute to the project and/or those with a significant interest in the outcome are a fundamental component of both VE and VM. Participants should be free to put up ideas and as far as possible idea generation and analysis should be kept separate. It is the role of the chairman to ensure that this is the case, and hence good facilitation skills and a measure of independence are essential characteristics of the role.

The process

Value techniques are founded on three principal themes:

- Achievement of tasks through involvement and teamwork; based on the premise that a team will almost always perform better than an individual.
- Using subjective judgement, which may or may not incorporate risk assessment.
- Value is a function of cost and utility in its broadest sense.

Key decisions in the application of value techniques are:

- When should the technique be utilised?
- Who should be involved?
- Who should perform the role of the facilitator?

A balance must be struck between early application before an adequate understanding of the problem and constraints has been achieved and late application when conclusions have been drawn and opinions hardened. Although feasibility (when identifying suitable options) and pre-construction (before design freeze) would in most cases be suitable, each project should be examined on its merits.

The facilitator's role is to gain commitment and motivate participants, draw out all views and ensure a fair hearing, select champions to take forward ideas generated and to keep to the agenda. To achieve these goals the facilitator must be independent, possess well-developed interpersonal and communication skills and be able to empathise with all participants. Although he or she must understand the nature of the project, this need not be at a detailed level. Large, complex or otherwise difficult projects may warrant employment of an external specialist facilitator. The facilitator's role is crucial to the success of the exercise and care needs to be taken over selection.

The purpose and the agenda for the value forum should be determined by the project team. A value statement should also be produced giving a definition of value in relation to the particular project. For example, value may not be related solely to cost but may also encompass risk, environmental impact, occupational utility, etc. Although the significance of these factors will be project specific, the project team must ensure that this statement reflects corporate policy. The statement is not intended to be a constraint but is used as a benchmark throughout the forum to maintain focus.

Two to three days are usually required for each value forum. The project manager must ensure that all supporting information is available to the forum in summarised form and that expert advice is readily available. The project manager must therefore ensure that personnel with a detailed knowledge of the project participate in the forum.

Link to risk assessment

Value techniques may be used in conjunction with risk assessment where there is a variety of means of managing risk and choices have to be made. The process is particularly valuable in identifying the optimum mitigation approach where risk management options impinge on a variety of project objectives.

In this case, risk management objectives are determined, in open forum, alongside overall project objectives. Risk management options are then ranked against the full range of objectives to determine the best option overall.

Potential pitfalls

- Cost (monetary and time) of value meetings can be high.

- At the feasibility/option identification stage, aspects of the technique can conflict with the principles of economic appraisal which seeks to identify an optimum solution by reference to an absolute measure of benefit as opposed to the subjective criteria used in value techniques, for example, a standard of protection could not be specified as a value objective. If benefits or costs could be assigned to all criteria, there would be no role for VM analysis at the feasibility stage, although the facilitative and team-building aspects of the technique would still be useful.

- Where economic appraisal overrides VM, team-building benefits will be undermined.

- Client participation in the value process could prove to be detrimental in the event of dispute with a consultant or a contractor. However, client participation is an integral part of the process, particularly at the feasibility stage, and must therefore be performed by experienced and knowledgeable staff aware of the contractual pitfalls.

Value techniques should be applied where there is a reasonable prospect of cost saving or substantial risk reduction or where consensus is necessary and difficult to achieve. Examples may include high value or complex projects impinging on a variety of interests and projects where social, economic, environmental and/or intangible benefits are significant but difficult to quantify. Value techniques may be used whenever there is a need to define objectives and find solutions.

An example of utilisation of VM at key stages in a project framework is shown next.

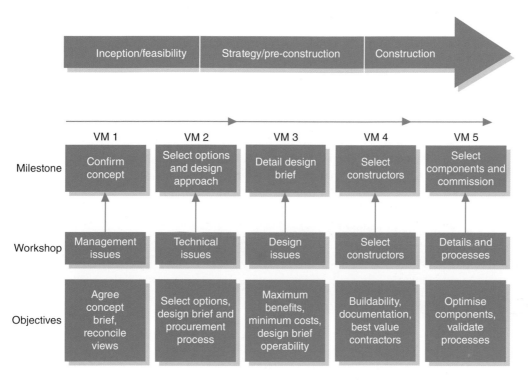

Stages of VM study.

Briefing Note 3.03   Project risk assessment

Risk is inherent in almost all construction projects. Depending on the nature and potential consequences of the uncertainties, measures are taken to tackle risks in various ways. Risk management is a systematic approach to identifying, measuring, analysing, and controlling areas or events with a potential for causing unwanted change. It is through risk management that risks to the schedule and/or project are assessed even though some risks are speculative and may bring about gains as well as losses.

A process of risk assessment and management has to be implemented at an early enough stage to have an impact on the decision-making during the development of the project. Small workshops should put together a list of events, which could occur, that threaten the assumptions of the project. A typical list requiring different workshop forum might be:

- Revenue

- planning consent

- schedule

- design

- procurement and construction

- maintenance and operation

Assessments should be made on:

- probability of occurrence (percentage)

- impact on cost–time–function (money–weeks–other)

- mitigation measures

- person responsible for managing the risk

- delete point where the risk will have passed (date)

- date for action by – transference – insurance – mitigation (date)

The mitigation measures need evaluating in terms of good value for money. Transference should be to the party most capable of controlling it since they will commercially price it the lowest. Lateral thinking might shrink, or in rare cases eliminate, risk. The risk register should be regularly reviewed and be part of all onward decision-making processes.

Risk register

The formal record for risk identification, assessment and control actions is the risk register. The risk register may be divided into three parts as follows:

- Generic risks: Risks that are inherent irrespective of the type or nature of the project.

- Specific risks: Those risks that are related to the particular project, perhaps identified through a risk workshop involving the project team.

- Residual risks: This is a list of risks identified above which cannot be excluded or avoided and contingency has to be provided for their mitigation.

The financial effect of such residual risks must be evaluated to determine an appropriate contingency allowance. A time contingency should also be considered.

When using the risk registers, values of occurrence and consequence are assessed using high, medium and low (H, M, L) values. Other forms of assessments include very high, high, medium, low and very low scoring or using numerical values (e.g. a ranking of 1–10, with 10 signifying a very high probability/impact risk and 1 identifying a risk of almost negligible impact/probability). An example of a part of a risk register is shown next.

Risk number	Description	Probability of occurrence (%)	Unmitigated impact			Mitigations adapted	Person responsible	Last updated	Delete point	Date to action
			Impact cost	Time	Function					
1										
1.1										
1.2										
1.3										
2										
2.1										
2.2										
2.3										
2.4										
2.5										
2.6										
2.7										
2.8										
2.9										
2.10										
2.11										
2.12										
2.13										
2.14										
2.15										
2.16										
2.17										
2.18										

Often risks are external arising from events one cannot influence. This particularly applies to markets and revenue projections. Some might even be 'show stoppers'. Value for money studies might throw up options of hedging or taking out special insurance. Obviously the risks of high probability and high impact are the ones on which to concentrate.

Contingency planning

Contingency planning is the development of a strategy to minimise the effects of intervening events that could possibly interfere with the smooth running of the project at some time between inception and completion. A contingency is a planned allotment of the time likely to be taken up by the occurrence of an intervening event.

Some scheduling software will offer a choice as to whether activities are, by default, to be scheduled as early as possible or as late as possible. Others default to one or the other. Where an activity would otherwise be scheduled as late as possible, the introduction of a contingency period buffering its end date will have the effect of scheduling the planned commencement of the activity earlier than would otherwise be the case. The effect of this will provide for a degree of delay in the completion of the activity to be absorbed by the contingency period. Designated non-work periods such as religious, industry-related or statutory holidays or weekends are not contingency periods and should not be treated as such. Only that party who is contractually liable for the consequences of the risk maturing can properly determine the quantity and distribution of the contingency it perceives to be required from time to time. Accordingly, contracts should (and generally do) make clear who is contractually liable for the consequences of the risk maturing and accordingly who owns the contingency.

In the same manner that cost budgets usually have an allocation of funding called 'a contingency sum' that the employer may rely on to spend against for unforeseen events, the schedule must have strategically placed contingency activities to absorb the time effect of intervening events that are at the employer's risk.

Prudent contractors will also make allowances for the risks they bear in the management and distribution of the resources and the quality of the work they carry out.

Contingencies should be designed to be identified separately for both the employer's and the contractor's risks and for those risks which are related to

- an activity, or chain of activities

- a contractor, subcontractor, supplier or other resource

- an access or egress date or date of possession or relinquishment of possession

- the works, any defined section, and any part of the works

At the lowest level of density, schedule contingencies are likely to be the longest in order to provide some accommodation for the unknown aspects of the schedule. Because of the absence of precision at this level of density, the separately allocated contingencies to one party or the other may both be arrived at by a formula adjustment. One way of identifying contingencies at this level is to use a formulaic approach such as Monte Carlo analysis to allocate an additional period to the known activities. The Monte Carlo algorithm randomly generated values for uncertain variables over and over again to generate model contingency periods.

At medium density, there is little scope in the schedule for notional formulaic calculations to accommodate unknown and unquantifiable risks and contingencies must be clearly allocated to one part or the other. There must be no contingency that is unallocated to an owner. At medium density, the risks should be clearly identified and a rational explanation set down in the method statement of the manner in which the possibility of the risk maturing has been allowed for.

At high density, the risks that need to be accounted for will be significantly fewer than at other densities. At this density contingencies must be clearly allocated to one party or the other. There must be no contingency that is unallocated to an owner and none clearly justified in the method statement. There may legitimately be risks such as inclement or adverse weather, unforeseeable ground conditions and utilities, plant breakdown, re-work, absenteeism which may need to be allowed for at this density but there should not be the need for design risk contingencies, or implied variations at this stage.

An example of a format of a risk mitigation table is shown next.

Risk mitigation table

RISK NAME:		DATE:	ISSUE No.:	ISSUE No.:
Risk category/Ref.:			Risk ownership:	

Risk evaluation		Probability	Cost	Time	(Other area e.g. environment/ health and safety)	Total score
		Current				
	Projected					

Risk description:

Risk mitigation plan: By whom: Review point/milestones:

End date/time scale

Comments

Project risk assessment checklist.

Project _____ Date _____

Overall risk assessment is:

Signatory	Signature	Date
Client		
Project manager		

☐ Normal risk

☐ High risk

Risk consideration	Criteria	Risk assessment Normal High				Proposed management of high risk

1 Project environment

Risk consideration	Criteria					Proposed management of high risk
User organisation	■ Stable/competent ■ Poor/unmotivated/untrained	☐	☐	☐	☐	▭
User management	■ Works as a team ■ Factions and conflicts	☐	☐	☐	☐	▭
Joint venture	■ Client's sole contractor ■ Third party involved	☐	☐	☐	☐	▭
Public visibility	■ Little or none ■ Significant and/or sensitive	☐	☐	☐	☐	▭
Number of project sites	■ 2 or less ■ 3 or more	☐	☐	☐	☐	▭
Impact on local environment	■ High ■ Low	☐	☐	☐	☐	▭

2 Project management

Risk consideration	Criteria					Proposed management of high risk
Executive management involvement	■ Active involvement ■ Limited participation	☐	☐	☐	☐	▭
User management experience	■ Strong project experience ■ Weak project experience	☐	☐	☐	☐	▭
User management participation	■ Active participation ■ Limited participation	☐	☐	☐	☐	▭
Project manager	■ Experienced/full-time ■ Unqualified/part-time	☐	☐	☐	☐	▭
Project management techniques	■ Effective techniques used ■ Ineffective or not applied	☐	☐	☐	☐	▭
Client's experience of project type	■ Client has prior experience ■ First for client	☐	☐	☐	☐	▭

3.03 Briefing Note

Risk consideration	Criteria	Risk assessment Normal High				Proposed management of high risk
3 Project characteristics						
Complexity	■ Reasonably straightforward ■ Pioneering/new areas	☐	☐	☐	☐	☐
Technology	■ Proven and accepted methods and products ■ Unproven or new	☐	☐	☐	☐	☐
Impact of failure	■ Minimal ■ Significant	☐	☐	☐	☐	☐
Degree of organisational change	■ Minimal ■ Significant	☐	☐	☐	☐	☐
Scope	■ Typical project phase or study ■ Unusual phase or study	☐	☐	☐	☐	☐
Foundation	■ First phase or continuation ■ Earlier work uncertain	☐	☐	☐	☐	☐
User acceptance	■ Project has strong support ■ Controversy over project	☐	☐	☐	☐	☐
Proposed time	■ Reasonable allowance for delay ■ Tight/rapid build-up	☐	☐	☐	☐	☐
Scheduled completion	■ Flexible with allowances ■ Absolute deadline	☐	☐	☐	☐	☐
Potential changes	■ Stable industry/client/ application ■ Dynamic industry/client/ application	☐	☐	☐	☐	☐
Work days (developer)	■ Less than 1000 ■ 1000 or more	☐	☐	☐	☐	☐
Cost–benefit analysis	■ Proven methods or not needed ■ Inappropriate approximations/ methods	☐	☐	☐	☐	☐
Hardware/software capacity	■ None or proven methods estimates ■ Unproven methods/ no contingency	☐	☐	☐	☐	☐
4 Project staffing						
User participation	■ Active participation ■ Limited participation	☐	☐	☐	☐	☐
Project supervision	■ Meets standards ■ Below standards	☐	☐	☐	☐	☐
Project team	■ Adequate skills/experience ■ Little relevant experience	☐	☐	☐	☐	☐

Risk consideration	Criteria	Risk assessment Normal High	Proposed management of high risk
5 Project costs			
Cost quotation	▪ Normal (i.e. time-based) ▪ Fixed price	☐ ☐ ☐ ☐	⬚
Cost estimate basis	▪ Detailed plan/proven method ▪ Inadequate plan/method	☐ ☐ ☐ ☐	⬚
Formal contract	▪ Non-standard form ▪ Standard form	☐ ☐ ☐ ☐	⬚

6 Other

Briefing Note 3.04   Information and communication technology

The success of construction companies is to a large extent contingent upon the success of their projects. Successful delivery of projects are highly data sensitive, reliant upon seamless exchange of heterogeneous data among a diversity of actors who may be dispersed geographically and in terms of their business models and systems. Inevitably, the industry has taken full advantage of the ICT albeit somewhat later than other industries such as manufacturing and aerospace industries. This section provides a brief account of some of the technologies that have been exploited to varying degrees, before homing on the details of a typical project management system. In order to avoid making assumption that IT is not just about technology, the section will commence by looking at process issues.

Business processes

A business process is 'a collection of related structured activities that produce a service or product that meets the needs of a client'. Business process management is a field of management that combines effectiveness and innovation to align organisations with the needs of clients.

The long established success of process approach in manufacturing promoted the notion of construction as a manufacturing process and paved the way for its recognition with construction as an effective way of increasing performance and consistency. However, it was with Egan (1998) that this view made its way to the forefront of business, policy and research community alike. Basically, a process is a sequence of steps to achieve an objective through integration of people, procedures and technology. It expedites business reengineering and provides a framework for continuous improvement. Subsequently, it lays the foundation for supply chain integration. While, process is a perspective, it also is a methodology. Cooper defines process as '... a formal blueprint, roadmap, template or thought process for driving a new product project from the idea stage through to market launch and beyond'. The adoption of this methodology by construction industry culminated in the development of the Construction Process Protocol developed at Salford University. Indeed, Construct IT's perspective on information technology is through construction processes.

Construct IT for Business is an industry-led non-profit making collaborative membership-based network, comprising leading edge organisations representative of the construction industry supply chain in addition to professional institutes and research and development (R&D)/academic institutions. Their aim is to improve industry performance through the innovative application of IT and act as a catalyst for academic and industrial collaborations. Their mission is 'To be an effective enabling and co-ordinating force (agent) in the application of IT within the construction process as a contribution to innovation and development of best practice'.

Interoperability

There are many independent and yet interrelated disciplines involved in the realisation of a construction project throughout its lifecycle. Each discipline is dependent on the use of its software tools that rely on input from applications in other discipline and in turn provide an output into them. These interactions are not always sequential and decisions may involve a number of iterations. Lack of compatibility of these application has created disjoints in the project development process as well as loss of information.

Interoperability has been defined by the Institute of Electrical and Electronics Engineers (IEEE) as 'ability of two or more systems or elements to exchange information and to use the information that has been exchanged'. Within the context of construction systems and processes, it invariably refers to seamless exchange of data among various software systems. Interoperability enables integration of systems within an organisation as well as those employed by their business partners. Subsequently, it has the ability to save time and money, as well as promoting collaborative working that yields several other advantages. By its nature, the implementation of interoperability requires development of standards. Interoperability can be achieved in various ways. However, by far the most comprehensive and consistent way of achieving interoperability is through Industry Foundation Class (IFC), which is an object-oriented data model class-based structure which facilitates sharing of information. IFCs provide the means for communication between applications. However, the use of IFC model server offers a more efficient way of data exchange in a multidisciplinary environment. These developments are a global effort that is coordinated by the IAI which started by 12 companies in September 1995 as a not-for-profit global alliance of construction and facilities management industries. IAI aimed to address representation of construction physical objects such as doors and walls as well as abstract objects such as process and space.

IAI Building SMART – UK Chapter specialises in establishing standards for the use of object technology in construction and facilities management. buildingSMART is the new branding of IAI started in 2005 specialised in using BIM and IFCs as the trigger to smarter ways of working. Its aim is to offer smarter ways of working which will directly affect the processes and skill sets used in the industry and other issues such as contracts, payment systems, insurance, education and training. It will require information links with business activities. They also aim to seek alliances with other similar motivated organisations which provide complementary standards and/or support processes which deliver faster, better, less expensive and more predictable results than can be achieved with traditional methods. Their mission is 'To provide a universal basis for information sharing and process improvement in the construction and facilities management industries'.

eBusiness

One definition of eBusiness ventured by Construct IT is '...as automated business processes (both intra- and inter-firm) over computer mediated networks and also the integration of all business activities that include redesigning of business process or reinventing of business model through information and communication technologies'.

eBusiness solutions are technologies for construction project life cycle collaboration. They are typically recognised as project extranet and software-as-a-service.

It is argued that successful implementation of eBusiness and e-commerce is contingent upon having a foundations of readiness. Furthermore, eBusiness framework needs to be complemented with business processes reengineering and change management.

The Network for Construction Collaboration Technology Providers (NCCTP) was formed in 2003 to bring the eBusiness providers together in a coordinated fashion.

It was anticipated that the network will help to promote the effective use of online technology to support collaborative working on projects and capital developments in UK constructions. It provides a single independent body with whom clients can communicate regarding the future development of collaboration technology. Their aim is to develop and implement an agreed set of data exchange standards between all members to enable bulk transfer of data from any one system to another. They also aim to develop and implement an agreed set of data exchange standards between all members to enable routine transfer of information between systems for cross project working. They will establish a group whose membership broadly represents the collaborative technology providers working within the construction industry and provide a vehicle to address generic market and technology issues. Their mission statement is 'NCCTP seeks to promote the benefits and use of collaborative technology in the construction and related industries'. Over the past few years, NCCTP has lost its momentum, though its infrastructure remains in place.

Electronic document management

Electronic document management systems (EDM) have proven to be a highly efficient way of exchanging and managing documents particularly within those industries where relationship between partners is long term. Despite extensive automation and digitisation, much of data and information communication during the design and build processes of construction is still paper based. This is partly due to contractual and legal reasons and partly due to the culture of the industry. Other than the consequential inefficiencies, the paper-based medium deprives reuse of the knowledge and lessons contained within these documents. The industry being highly data intensive, the concept of document management has developed some interest within the industry. The low satisfaction with the solutions offered by vendors has led some larger companies to develop their own document management system. These systems have been criticised for not being user friendly and for being difficult to integrate with other systems such as CAD. There are also security issues associated with protecting and ownership of the data. But the main pitfall of EDM has been the upfront work required to setting up the system for all participating partners. Indeed, every partner within the project must use the same EDM system on a project in order to be able to access and share documents. Also, due to a lack of attention to semantics of the information being processed, the facilities offered are often limited to document storage, retrieval, versioning and approval.

Electronic trading

Electronic trading is an attempt to minimise paper transactions. It is about sending and receiving transactional data either directly or through a hub, the role of which is to apply the rules of connectivity. Depending on the nature of the business and choice, there may be more than one hub.

It is evident that considerable cost and time saving can be achieved through the implementation of eTrading. Additional advantages include increased visibility, control and certainty. There are a large number of companies connected for eInvoicing. A study of actual customer case has shown that a 57% saving can be achieved through electronic automation of simple invoicing. While electronic trading is applicable to transactions throughout the life cycle of construction processes that involved collaborative working, it tends to offer greater advantages in areas involving tenders, requisitions, orders, invoices, acknowledgement, delivery, statements and remittance.

Construction Industry Trading Electronically (CITE) is a collaborative electronic business initiative for the UK construction industry where data exchange specifications are developed by the industry for the industry, enabling the industry to move forward together. It was launched in 1995 and formally marked the start of a major collaborative undertaking with active participants from the professions, contractors, sub-contractors and suppliers. The overriding aim of CITE is to extend the operational use of electronic business across the construction industry in the widest context and by doing so create an open trading environment for all. Where extending services or standards apply, CITE seeks to adopt these and build on best practice. Their mission is 'To develop and promote the adoption of eBusiness standards in the construction and facilities management industries'.

Enterprise resource planning

Enterprise resource planning (ERP) is an enterprise-wide information system that helps an organisation to integrate, optimise and manage its various functions ranging from product planning, material purchasing, inventory control, and product distribution to taking orders, finance, accounting and human resource management. It uses one application, one database and a unified interface across the whole enterprise. While the concepts of Inventory Management & Control System, Materials Requirement Planning and Manufacturing Requirements Planning date back to 1960s, 1970s and 1980s respectively, it was during the 1990s that ERP started to play an important role in integrating the information systems of the whole enterprise, thus enabling efficient management and use of resources including human, materials, finance and information. ERP systems are designed to underpin both process-oriented aspects of the business as well as business processes that are standardised across the enterprise. Despite their slow start, their use within the construction industry has been on the increase.

Concurrent engineering

Concurrent engineering is another concept adopted from manufacturing industries in order to reduce cost and time as well as to improve the overall quality of the product. Concurrent engineering helps to minimise ambiguity and unforeseen issues relating to the design, as it allows consideration of the downstream product life cycle issues during the early design stages. This feature applies through the life cycle. In construction, the concept has embraced detail design, environmental impact, space planning, maintenance, operational issues, emergency evacuation, security and construction. Subsequently, concurrent engineering has facilitated collaborative working of many actors including architects and designers, contractors, planners, engineers and maintenance engineers. This collaborative environment is further aided by virtual reality (VR) and communication tools. More advanced collaborative tools have enabled distributed collaboration of geographically distributed parties.

Information visualisation

Human vision and domain expertise are powerful tools that (together with computational tools) make it possible to turn large heterogeneous data volumes into information (interpreted data) and, subsequently, into knowledge (understanding derived from integrating information). Visualisation can encompass a variety of topics ranging from information visualisation (IV) to scientific visualisation, virtual reality, multimedia, etc. In this sub-module, the issues relating to visualisation will be grouped under three sections. This relatively new field of science is gaining significant importance, because this field has become rich in tools and techniques within a wider multidisciplinary scope.

Information visualisation cannot be defined precisely, nor there is a consensus as to what constitutes IV. Some argue that IV should involve the use of computers, but others simply argue that it is any form of visual tool that helps us to develop a better understanding about a concept that has underlying data. There is also the third view that suggests that information visualisation relates only to non-numerical data. This notion has been supported by some innovative work. Certainly, there are several landmark examples supporting the claim the IV is akin to non-numeric situations. Probably, a good example of this is the map of London Underground.

Since the development of IV as an independent and credible field of science, there has been a considerable move towards introducing new techniques. Especially, with the increasing role of the internet and World Wide Web (WWW) as the source of distributed information across the globe, the need for the development of new tools has become apparent. Another reason giving rise to the emergence of IV tools is due to the increasing size and complexity of today's data. Subsequently, it has become evident that such complex data could not be expressed or visualised through the use of traditional visualisation tools. This is particularly relevant because the true purpose of visualisation is to develop better understanding about the data and the concepts they represent. These issues have, therefore, given rise to the development of several tailor-made IV tools – that is tools that are designed especially for the data and the users of the data. Examples of techniques include Glyph World, TreeMap, Cone Tree, Arches, Hyperbolic, Landscape Tree and Websome Map.

Mobile technology

Mobile technologies have been used to in various ways within construction industry. They have made considerable inroad towards eliminating paper works associated with numerous processes within this data-intensive industry. The real benefit of mobile technology is indeed its inherent characterise of being mobile thus exploitable anywhere anytime. This feature, in conjunction with the fact that all of its transactions are in digital format, makes mobile technology an imperative part of construction management. Most popular are the types of robust personal digital assistant (PDA) and rugged Tablet PC devices for semi- or fully-automated data collection in an structured way so they can be synchronised with the back-office systems via WLAN or GPRS. The ability to communication upstream allows operatives with a PDA to access necessary distributed helpful data/information. Issuing orders, goods received numbers (GRNs), quality inspection reports, method statements, and RFIs are just a few examples where operatives on site can communicate with relevant stakeholders or systems to archive, take actions or check compliant.

PDAs are becoming more sophisticated in terms of both hardware and software technologies. The latter includes devices that are capable of receiving updated revisions of drawings to the site and return updated information. This is just as true about documents such as drawings as is for data relating to project progress which can be captured by the site engineer and synchronised with the master project management system or a project integrated database. Applications have been developed for mobile devices to undertake specific tasks. For instance, the mobile manager by Primavera allows interaction with the main Primavera project management software. In effect, mobile technologies become extension of main systems.

PDA devices or tablets can be enhanced or complemented with Radio-frequency identification (RFID) readers, and thus used for inspection and tracking and if necessary to issue hazards notification (e.g. for equipment misuse or prolonged duration of exposure to hazardous materials or equipment) and issue automatic alerts via email or SMS. The sharper end of the technology can take advantage of camera, smart hard hats and even augmented reality for the identification of maintenance equipment's, material and operatives. The latter could be with the aid of operatives' mobile phones.

The innovative exploitation of mobile technology in the UK construction industry is coordinated by Construction Opportunities for Mobile IT (COMIT). They have identified several areas where mobile technology can be exploited within construction industry. These include the following areas for which practical solutions are offered, and where necessary their process maps are generated:

- Health and safety:
  - Collection of audit/inspection data of the field
  - Notification hazards to be remedied
  - Retrieving the training records of operatives
- Drawing distribution and usage:
  - Delivery of drawings to site
  - Notification of revisions to drawings
  - Capturing of as-built information
- Goods received notes:
  - GRN creation and distribution
  - Tracking of delivered goods
- Maintenance inspection:
  - Delivery of work orders to the field
  - Collection of maintenance information
  - Locating spare parts and equipment
- Monitor hazardous activities:
  - PDA form for recording time spent undertaking hazardous activities
  - Automatic identification of hazardous activities and operatives
- Monitor progress:
  - PDA form for capturing progress made
  - Integrating with back-end systems
  - Benefits of mobilisation
- Onsite resource monitoring
- Quality inspection
- Site design issues resolution
- Site records
- Task allocation:
  - Delivery of method statements to site
  - Capturing of briefing acknowledgements
  - Capturing of progress information

COMIT started as a two-year research and development project (from August 2003 to December 2005) partially funded by the Department of Trade and Industry (DTI).

The project brought together representatives from construction, technology, research and dissemination organisations to form the COMIT community. In September 2005, a self-funded organisation was established. COMIT's aim is to help its members to realise benefits from the use of mobile ICT. Their mission is 'To become the European centre of Excellence for the exploration, development and implementation of emerging ICT within the construction industry'.

4D project management

An alternative approach to construction planning and project management is to develop a visual representation of the building and manage the construction process by visualising the process and the product at any point in time. As a complementary tool to the traditional project management tools, the visual approach yields additional capabilities: Visual inspection of clashes, the identification of the spaces needed for resources at any time and access to a common database are a few of many examples. In this environment, project participants can increase productivity and reduce waste by visualising and analysing construction schedule and communicating in an environment that is conducive to collaborative working.

The potential benefits of the visualisation of construction progress as an alternative or complementary method to the traditional methods have been realised as early as early 1990s. Indeed, the first generation of such provisions commenced in 1987 by Bechtel and Hitachi Ltd. In United Kingdom, a handful of companies began to focus their business on 4D construction visualisation, initially offering a visual representation of the building before committing the client to larger expenses of providing a time-based visual representation of the construction schedule progress. The methodology relies on the combination of the traditional project management tools with a CAD-based visualisation tool. Further enhancements have been achieved through the introduction of decision support systems for the planning process. In some ways, these efforts laid the foundation for what was later know as BIM. In particular, the opportunity for collaborative working of the main parties engaged in the design and construction phases.

In addition to encompassing customer relationship management, financial and budgetary management including cash flow forecasting, property and asset management and project scheduling, visualisation tools have made a significant impact in delivering project management at a global scale.

Application of project scheduling software

At its lowest level, the software sold for scheduling as 'project management software' can be no more than a drawing tool or, at its highest, a complex arrangement of customisable relational databases and graphical front end. In order to be capable of producing a schedule that can perform as a time model, the software must have an adequately functional relational database at its core. The reason for this is that the software has to be capable of computing the consequences of change, while a drawing tool, which simply illustrates the decisions made by the drafter, cannot do this.

Many schedulers are trained by the software manufacturers to operate the software they use. This is extremely important and useful training. However, this is not intended to increase competency in terms of time or schedule management within the project environment and must not be mistaken to be a substitute for these important skills. By analogy, many of us have experience of securing a good grounding on how Microsoft Word works, but even with its spelling and grammar checking option, the software will not guarantee that what is written is useful, technically accurate or even intelligible.

No matter how high the quality of the software, it cannot produce a high-quality output of its own accord. Even the best project planning software will not secure the competent management of time.

While every company considering software products will from time to time wish to take into consideration matters peculiar to themselves, or matters peculiar to the project on which they wish to work, there are certain considerations that should transcend subjective preferences and there are certain software attributes which are desirable for the purposes of competent time management. Because software changes by the day and 'new and improved' products (which unfortunately is a term often confused with 'more bells and whistles') are brought to the market, those attributes that are desirable only for the purposes of time management, irrespective of whether they are currently available in any particular product, are listed below.

It is unhelpful if different parties to a particular project use different software because different products work in different ways and even if given the same data will produce different calculations from different algorithms. Accordingly, all parties to a project should use the same software and a departure should not be permitted.

While getting to grips with unfamiliar software may be tedious, competent schedulers will generally know their way around any new product within an hour and will rarely take more than a day or two to become sufficiently capable of using it competently. Unfamiliarity with scheduling software products should thus not be a serious consideration in product selection.

Primary software considerations

Projects and subprojects

Software that can only cope with a single project at a time is unlikely to be sufficiently flexible for complex projects. For example, there may be necessities to identify individual key dates, sectional completions or indeed separate operational zones as subprojects, for ease of application in practice, it may also be useful to identify separate operational zones as subprojects.

Activities

For each activity there should be (a) a unique activity-identifying alphanumeric code and (b) a unique description. Software that permits duplication of activity identifiers (IDs) or activity descriptions without warning is likely to produce schedules that lack clarity and are thus incompatible with good practice. The software should not facilitate duplication at all, or, if it does, should have a clear permanent warning on the schedule as to the deficiency.

The software should be capable of distinguishing between the following activity and event types:

- duration-identified activities
- resource-calculated activities
- hammocks
- start milestones or flags
- finish milestones or flags
- employer-owned contingency/risk allocation
- contractor-owned contingency/risk allocation.

An activity-related field capable of taking free-text and numbers as comments or notes is often a useful facility. The software should be capable of identifying activity durations in different formats. Although for most purposes in construction, activity durations in days may be sufficient, for the purposes of limited possessions, durations in hours and minutes, and in outline schedules, durations in weeks and months are necessary. The software should make it clear to what unit of time it carries out its calculations, that is days, hours, minutes or seconds. The best software calculates to the minute.

The software ought to be capable of identifying which activities are logically determined to be of a shorter duration than the applied logic and whether they are to be 'stretched' or 'not continuous' as a result of the logic, or the logic changed.

Logical relationships

The software should permit a logical flow of work and prohibit the indication of relationships which are impossible to perform. Any software that fails to do that is likely to produce schedules that are incompatible with good practice and should not facilitate it at all or declare a clear permanent warning on the schedule as to the deficiency.

The software should be capable of identifying all variations of logical links either individually or in combination. Software that limits the user to finish-to-start logic or few logical connections to any one activity is unlikely to be useful.

The software should identify any inconsistency between logic and the activity durations to which the logic is applied. Logic should be capable of being illustrated as 'driving' or 'non-driving' to any chosen point within the model. Logic should distinguish between the following:

- engineering logic (the construction sequence with no resource constraints)

- resource logic (the construction sequence carried out with the available resources)

- preferential logic (the construction sequence with imposed constraints to modify the purely 'engineered' and/or 'resourced' construction sequence)

- logic linking zones and or subprojects.

The software should be capable of identifying fixed lead and lag, as well as the working calendar that the lead or lag schedule is required to adopt. Lead and lags should be listed as logic attributes.

Constraints

Manually applied constraints are likely to be useful on most projects. Those which are acceptable, when correctly applied are as follows:

- start no earlier than a given date

- start no later than a given date

- start as late as possible (also known as zero-free float).

The software should be capable of clearly identifying when a manual constraint has been applied to an activity. Some software facilitates the use of constraints that will manipulate criticality and inhibit the ability of the software to accurately model time. These are seldom acceptable in a schedule used to manage time and which, if available in a software product, should not be permitted to be used without a clear permanent warning on the schedule as to their effect. Possible constraints are:

- any combination of constraints that will fix the earliest and latest dates for any activity or milestone

- a mandatory start date

- a mandatory finish date

- zero-total float.

Critical path

The software should be capable of identifying:

- the longest path to completion

- the longest path to intermediate key dates or sectional completion dates

- logic and activities that are critical from those that are not critical to one or more completion dates

- total float on each path

- free float on each activity on each path.

The software should be capable of facilitating the tracing of a critical path or paths through the driving logic of each activity on the critical path to a particular completion date or key date from time to time.

Calendars

The software should be capable of facilitating the use of a number of different working calendars for activities, resources and lags, each capable of identifying:

- working week start day

- working weeks and weekends

- working days

- working hours

- holidays

- standard calendars and exceptions.

Resources

The software should be capable of facilitating the use of a number of different resources and determining whether or not the duration of the activity to which they are allocated is to be calculated by reference to the specified resources in terms of:

- name of resource

- unit working period

- allocated working calendar

- number of units of work per calendar period

- cost per unit

- availability.

Work breakdown structure and activity coding

The software should be capable of identifying a work breakdown structure. While a structure of eight levels should be the ideal, a structure of less than five levels is unlikely to be practical on a complex project. The facility for a broad variety of bespoke database fields that can be displayed is usually an essential requirement of complex schedules.

Organisation

The software should be capable of organising the layout in any combination of fields and attributes, sorting on activity and logic and attributes and values in any field.

Filtering

The software should be capable of filtering the content of any layout to select the value of any field, or attribute, either alone or in combination with other fields, or attributes on the basis of the value or attribute:

- equal to
- containing
- not equal to
- not containing

and where the field contains a value, there should also be the facility in relation to those values to satisfy the logic test of the value being within a certain range or not being within a certain range (i.e. and true or false response to a logical query (lower limit) < (value) < (upper limit)).

Layout

The minimum available layouts should comprise the following:

- bar chart without logic
- bar chart with logic
- network diagram (Arrow diagramming method (ADM) or Precedence diagramming method (PDM))
- resource profile
- cost profile.

The software should have the facility for creating and saving a variety of different combinations of fields and attributes organised and filtered, as layouts for reporting purposes. The time scale to which the layout is restricted to view should be identifiable to any duration and density during the period between six months prior to inception of the earliest project and 12 years after planned completion of the latest project.

Every layout should be printable as both hard copy and a portable document format (PDF).

As-built data

The software should be capable of identifying the factual data for each activity and resource as:

- actual duration
- start date
- finish date
- percentage complete
- remaining duration
- calculated cost

- actual cost
- certified value
- resources expended.

Updating

The software must be capable of identifying a data date by a straight line through the activity bars at that date. The software must be capable of comparing progress against the currently agreed baseline, such that any delays and/or changes in activity sequencing are clearly demonstrated.

The software must be capable of recalculating the critical path or paths and the predicted early and late start and finish dates of all activities and resources against the data date with the effect that:

- all activities indicated to have started or finished are indicated to have started or finished earlier than the data date
- no activity is identified to have started or finished later than the data date
- activities that are in progress at the data date are indicated to be due to finish on a date after the data date proportionate to their degree of progress in relation to their planned duration at the data date.

Inputting and editing data

The software should be capable of holding input data and edits in memory so that they are subject to undo and only saved on a positive instruction to do so.

Archiving

Files should be capable of being saved in compressed data format for archival purposes.

Training and support

The availability of effective, product-related training is extremely useful even for experienced schedulers. Even with the simplest of software, it is always helpful to understand what the software supplier identifies as the way it should be used. Because of the sophistication of modern software and the inability, or unwillingness, of the manufacturers to subject products to rigorous testing before release, committed and easily available software support and continuous updated software is more important today than it has ever been.

Secondary software considerations

Those matters which do not add to the quality of the calculated output but to the manner of use and which, depending on circumstances, may be of some importance are listed below.

Enterprise-wide software

Enterprise-wide software can directly link a project or projects across the internet such that large, complex projects are able of being scheduled and effectively monitored across the world. This software can relate all projects together with which the company is concerned. It is a useful attribute for enhancing company management.

Communications

Considerations have to be given whether the schedule may be accessed by other parties, in whole or part, with the appropriate access limitations (for example read only, read and edit, read and restricted edit etc.) as pre-defined and controlled access rights will be of importance in managing the schedule.

Appearance

Software capable of being customised according to the company requirements for house style by using different fonts, line thickness or type and colours for each available field or value in the database and the background is useful.

A drawing facility that can be used to highlight aspects of a report is also often useful.

Comparison of schedules

For the purpose of identifying the effect of differences between schedules in the process of review, revision, updating and impacting causative events, it is useful to have the facility for comparing two or more schedules on a line-by-line basis. In practice this usually means the facility for identifying one or more target schedules that can be viewed simultaneously with the current schedule.

Organisation

A facility for organising the layout in order of logical predecessors and successors is useful.

Transparency with other software

The facility for importing from and exporting to other scheduling software may be available, but if it is, it should be capable of listing the differences that result from such import or export.

Integration with time and cost-keeping systems can facilitate automatic updating from time, plant and material records which, in relation to a fully resourced schedule, can produce an automated update facility.

The facility for attaching hyperlinks to activity IDs should be available to facilitate the linking of such documents as photographs and movies, flow charts, procedures and method statements and progress records.

Risk analysis

The facility for stepping through a potential shift in timing of activities to ascertain the consequent shift in the critical path is useful. Monte Carlo analysis will give a profile of likelihood of success against given criteria which, if accurately predicted against data that remains unchanged, will predict likely outcome.

Archiving

A backup capable of being set to default periods or to be executed manually is a useful facility.

Briefing Note 3.05   Building information modelling

BIM definition

> 'Building Information Modelling is digital representation of physical and functional characteristics of a facility creating a shared knowledge resource for information about it forming a reliable basis for decisions during its life cycle, from earliest conception to demolition.'
>
> This definition by the Construction Project Information committee (CPIc) is based closely on the US National BIM Standards Committee (NBIMS).

Notice there is no mention of 3D modelling! The 'Building' in BIM might be a built environment asset of some kind, for example, a piece of infrastructure, a nuclear power station, a rail station, airport, a suburb of a city even or simply just a building. The asset has a lifecycle, which will follow the time line of definition, procurement, delivery and operation and eventual recycling.

The project BIM is the database of information about that asset that potentially can live with it over its entire lifecycle. From briefing information, through design, procurement, costing, contract, site delivery and construction, through to completion, handover, and operations and maintenance information to form the basis of FM services. And then subsequently provides the basis for re-use, adaption, extension, to eventual demolition, removal and recycling.

The information in that database could be in many formats. Yes there will be 3D design files, say in Revit, Bentley or Tekla, but there will also be spreadsheets, word documents, pdfs and all kinds of files containing technical, programme, cost information and any other information that can be attached to the database.

Current thinking seems to moving to the concept of having a relatively small model file, linked to a relational database in SQL or something similar which holds the majority of the information. Naturally this information can be catalogued, searched, queried and output as required.

So while many people equate BIM merely with 3D modelling, it is in actual fact much more to do with *data management over the asset lifecycle*. If we imagine all the information that is collected and circulated over the lifetime of a project – BIM enables an environment where that information and knowledge can be integrated and searchable.

BIM itself is not new technology – it has been used in other industries and sectors for several decades. Forerunners of the current BIM technologies were being used in the 1980s.

At this point, it is worth mentioning COBie – the Construction Operations Building information exchange. This is a specification or format for the capture and delivery of information needed by facility managers over the asset lifecycle. COBie information can be produced at all stages of the project and currently is usually presented as an extensive spreadsheet. Part of the UK Government BIM strategy is for public sector departments to determine their required COBie drops or data exchanges at each of the project stages. These guidelines, including illustrative examples and templates, have been published on the Government Task group website as COBie UK 2012. (*See resources given later.*)

Aspects of BIM

Using BIM enables a number of aspects and activities as follows: naturally this is reasonably neutral in terms of mentioning particular software or packages or companies facilitating BIM process or activities.

3D design coordination and clash detection

This is probably one of the most well-known aspects. Taking each of the discipline design models, it is possible to aggregate the models into one environment and identify clashes and conflicts between the discipline designs. Whether this is one totally integrated model or a 'federated' model composed of several models is a discussion. Certainly the federated approach reduces file sizes and lessens the impact on ICT infrastructure, which is a big consideration in implementation. It is possible to federate models produced by different software, through the use of packages such as Navisworks or Solibri, which will enable clash detection.

It is also possible to use Industry Foundation Classes (IFCs) to import a model into another package. There are varying reports on how successful this can be, but certainly the future will be about more interoperability between platforms, moving to an 'OpenBIM' environment.

4D timeline/programming

This is the generally accepted '4D'. It is possible to assign time attributes to parts of the model so that a view can be generated which shows the construction sequence. Into this can be added elements such as cranes, hoardings, scaffolding, hoists, etc. Most of us have probably seen these animations. Generally these are used to explain the construction sequence and site logistics to say, a client, site team or supply chain. This is particularly useful for understanding the requirements for the construction sequence and the implications in 3D.

5D quantity take off and cost planning

Once the model is designed, even in outline, it is possible take off quantities for elements or components. The output can be in a form which cost planning or estimating software can accept, although how this actually works needs to be defined upfront to make the technical aspects feasible. Over time as libraries of project-based data are established, this will enable information to be accessed to compile cost plans quickly and efficiently.

It is worth noting that the use of libraries of information to populate the model and database are a great factor in improving efficiency, speed and reliability of the project process.

Simulations – lighting, fire, people movement, thermal, carbon, energy

In addition to the basic 3D design packages, there are additional plug-ins and various packages that can use the model to simulate various conditions such as lighting levels, fire performance, the movement of people, including lift simulations and queuing patterns, thermal losses and transmittance, carbon emissions and energy consumption.

In compiling the model, the relevant data will need to be input as attributes to elements or components to enable the functionality required. But again as libraries of data are developed, then the input activity can be accelerated.

Operations/maintenance (O/M) manuals and information

The project BIM can be used as a means of storing and accessing O/M and CDM Health and Safety File information. This can be made available to a hand-held tablet, so by going into a room, you could call up the room data sheets, relevant manufacturer's information and so on. If say an item of plant or furniture, fixtures and equipment (FFE) has a barcode or an RFID tag, then this can be scanned and used to access the information on that item, such as dates of manufacture, installation, defects period, maintenance and any other related information you care to choose to have stored.

Visualisations

Another aspect that is probably familiar to most people. Visualisations of the design model, utilising stills from selected 'camera' views or animated fly-throughs around the site and building. These are particularly useful in explaining the scheme to customers or stakeholders. Also for explaining the scheme to the planning department or other interested parties. It is possible to have a fully virtual immersive experience, with headsets, goggles and gloves, so that you experience the model from inside. Most of the BIM-enabled universities such as Salford have this kind of facility.

These kinds of animations are extended into the construction sequence animation discussed earlier.

Site safety planning

Some contractors use the phrase 'Build it twice' that is, build the project virtually, and then go and physically build it. As the BIM model becomes more detailed and contains the construction status information, then it is possible to review the construction process in detail from a safety perspective. Considering sequences, access requirements, particular areas that may require special temporary works or other factors.

This takes risk assessment to a new level, enabling teams to view how to construct virtually, before going out into the field to actually do it.

Fittings, fixtures and equipment

Having designed the 'base-build', it is equally straightforward to design in FFE requirements, link these to a room data sheet package such as Codebook, and link to procurement, O/M information, etc.

Manufacturers are beginning to produce libraries of their product ranges, which will make it easier for designers to incorporate them in their models.

There are library sites being developed such as NBS BIM library and the BIM store.

Progress management

Using the time line and logistics elements, it is possible to translate the proposed programme of construction events into an animated view. Site managers using tablets can record progress and problems as they walk round the site and these can be uploaded to the BIM to give a model view of progress.

There is an interesting study called KanBIM which looks at BIM-based progress combined with lean thinking.[2]

Off-site manufacture

Having developed a detailed clash-free, data model, subcontractors can retrieve the information they need to manufacture their elements and components. In some cases, the data can be exported straight to CNC-enabled plants, which enable computer-generated manufacturing processes. Double or triple handling of information is avoided, and information is taken direct from the model to facilitate the process. As supply chain develop object libraries of their systems and components, these can be made available to designers to incorporate within the design stage BIM. Of course with the physical data can be added other attributes such as procurement data, costings, lead in times, technical performance data to enable simulations and O/M information.

Lifecycle costing and management

Using plug-ins or additional software packages, it is possible to run simulations of the lifecycle, looking at carbon emissions, energy consumption and so on. Therefore, from an early point in the design process, decisions can be made from an operational perspective to develop the optimum solutions.

Facilities management/building operations

The project BIM is updated to as-built at construction completion. Loaded with the O/M information, this can then be used by the customer's FM team to manage the asset.

This aspect is a key part of the UK government's drive on BIM adoption for the industry. If consistency of data is applied across the public sector estate, then this can be harnessed to result in increased efficiency and management.

As discussed earlier, the BIM can be accessed using hand-held tablets so that managers walking round the asset can retrieve and also update information. The use of barcode readers and scanners can enable quick access to relevant information on particular items of plant or equipment.

The use of tablets has been extended into snagging and defect logging following similar principles.

Recycling

At eventual demolition and/or recycling, the BIM can contain the information to enable safe demolition and also the material specifications for recycling if required.

In addition, assuming the BIM is kept up to date then should alterations or extensions to the asset be proposed, then the BIM provides a good starting point for the new design team.

RFID (radio frequency identity tag)

These really are intelligent barcodes. A small chip can be fixed or embedded in a component or object. This can be then read by a radio frequency reader, which detects the barcode number, and can be used by the database to identify the object and call up any related information.

[2]   http://www.tekla.com/international/about-us/news/pages/bims-return-on-investment-wheres-the-beef-now.aspx (accessed March 2014).

One application is that through the use of scanners at site entrances, contractors can track deliveries of materials, which could link to the progress-tracking element of a project BIM.

Items of plant or furniture or other equipment can have unique IDs, which enable the correct and relevant information to be called up from the BIM when scanned, as used by FM managers.

Refurbishment/retrofit

Much of the progress on BIM has been made on new-build, but now survey models are more readily available through the use of laser scanning technology.

Through the use of laser scanning stations, survey data is collected and converted into a 3D model to be imported into the BIM model. This provides greater accuracy earlier in the design process and is also on a par with traditional survey methods costs. Incidentally the same technology is being used to scan road accident scenes, speeding up the process of re-opening the roads to traffic.

Compliance/validation

An advantage of working in the BIM environment is the ability to check and validate various aspects of the model geometry and datasets at any given stage.

Evaluation tools can be used to check completeness and accuracy of information, at any stage, which gives teams much more certainty in decision making. Models can be checked for planning and Building Regulation compliance using rule checking platforms. Contractors can very quickly evaluate received tender information to improve the tendering process. The ability to evaluate information, gives the overall process transparency, more accountability, improves collaboration and team working.

A planning permission online tool has been used in Singapore for over a decade – decisions are made within a day of submission.

Implementing BIM

BIM is much more than technology and much more than just 3D geometric modelling. It opens up a different way of working, providing reliable data to across the asset lifecycle to the whole project team and stakeholders.

The more difficult aspects of BIM to deal with are intertwined in process, culture and team roles. The dynamics of BIM change the way project process works and the flow of information.

Previously we were in PUSH* mode, as information is produced, the outputs are sent out to those that need them, for example, drawings and info to subcontractors to start CDP design. In BIM, the model is effectively the output, and if information is needed by someone they just go and get it, when they need it, and define how they want it, in terms of format, presentation, etc. In BIM, the responsibility shifts to PULL* – those that need information for their activities get it from the source – the project BIM.

Benefits

The benefits of working in BIM are now well catalogued. The BSI Investors Report provides a good snapshot of key issues. The BIM Handbook catalogues a number of detailed case studies. (Both referenced later.)

---

* PUSH & PULL: A push/pull system in business relates to the movement of information or product between two subjects.

BIM offers a number of benefits including reducing the rework of design and site works, resulting in more certainty in the design and procurement process, at an earlier stage. This enables more efficient ways of working, as well as producing more viable design solutions that are value managed closer to a customer's requirements right from the start.

This can result in significant time and cost savings over the asset lifecycle as well as in just the design/procure stages.

In discussions with designers, they are open about producing better work, using fewer resources and producing information faster. A contractor publicised recently that for a modest investment on BIM on a new project, they had more than recouped their investment in a few months, through clash detection and avoidance, and risk elimination.

UK BIM strategy

The UK BIM strategy and implementation is driven by the Government Construction strategy developed by the then Chief Construction Adviser, Paul Morrell, and his team.

There are two main thrusts to the strategy:

• to reduce capital expenditure in the public sector by 20%

• to reduce carbon emissions, meeting the climate change targets

The long-term impact of the BIM strategy is to reduce lifecycle costs and increase efficiency of the procurement and management of the public sector estate.

Various groups and working groups were set up and are now producing standards and guidance documents for the UK industry. Pilot projects are also being run, sponsored by several government departments.

For more information, refer to the government BIM website[3]

Resources

Books/publications

BSI Investors Report, (free download from UK Government website)

• 10 truths about BIM – WSP, (available free from http://www.wspgroup.com)

• BIS BIM Report, (free download from UK Government website)

• BIM Handbook, 2nd edition, Eastman, Teicholz, Sacks, Liston, Wiley, 2011

• The impact of Building Information Modelling, Ray Crotty, Spon Press, 2012

• BIM Demystified, Steve Race, RIBA Publishing, 2012

Websites

www.bimtaskgroup.org *UK Government BIM website*

• www.cpic.org.uk *Construction Project Information Committee*

• www.bimgateway.co.uk *Classification and information*

• BSI B555 Road Map, *BIM maturity levels and standards information* http://www.bsigroup.com/en/sectorsandservices/Forms/BIM-reports/

---

[3] http://www.bimtaskgroup.org (accessed April 2014).

- http://www.buildingsmart.org.uk *BuildingSMART UK chapter*

- https://www.bsria.co.uk/services/design/soft-landings/ *BSRIA Soft Landings and handover/commissioning process*

'BIG LAUNCH' BIM documents – February 2013

The UK BIM Taskgroup and its working parties have been working on a number of key documents over the last two years.

These have been issued and all are freely available at the Government BIM website. The Task Group on the documents welcomes feedback as they are used by industry.

Just register and log in at http://www.bimtaskgroup.org/task-group-labs/

So, the documents are:

Government soft landings

http://www.bimtaskgroup.org/gsl/

'To champion better outcomes for our built assets during the design & construction stages through Government Soft Landings (GSL) powered by a Building Information Model (BIM) to ensure value is achieved in the operational lifecycle of an asset.' – Task Group website

'BIM+GSL=Better outcomes'

Based on the BSRIA Soft Landings process, GSL encourages the engagement of the project end users right from the start of design for any built asset. This improves the built asset design, construction and operation process.

The Golden Thread of GSL runs from the start of a project, linking, clients, end users, designers and constructors, focusing on outcomes and operational performance.

Digital plan of work

http://www.bimtaskgroup.org/digital-plans-of-work/

A working group has looked at the plethora of industry plans of work and produced the digital plan of work (dPoW). This provides a harmonised stage structure which will provide an overarching framework for all other plans of work produced by the Institutes such as the RIBA Plan of Work update due later this year.

Included in the dPoW are activities required for each stage and links with COBie requirements and the Employers Information Requirements.

Data hierarchy

http://www.bimtaskgroup.org/data-hierarchy-overview/

Linking with the dPoW, the Data Hierarchy defines the information requirements for each stage from general to detailed, including the Coordinated Work Stages, the Plain Language Questions which set out what information is required and also the Demand Matrix which sets out the information to be included in the COBie file that forms part of each of the information exchanges in line with the dPow.

Uniclass 2

http://www.bimtaskgroup.org/uniclass-2/

3.05 Briefing Note

For a digital information environment like BIM, we need a digital classification system, which is Uniclass 2. This is still effectively a beta version, but development is ongoing, and the classification will continue to evolve with our increasing use of BIM. The classification not only needs to be capable of developing with the growing data maturity of a model, but must also accommodate changes over the asset lifecycle and be capable of use by all stakeholders in the process.

COBie tools and testing

http://www.bimtaskgroup.org/cobie-tools-and-testing-overview/

Guidance on COBie UK 2012 has been available for sometime, but in this update an example project is modelled at various stages with corresponding COBie outputs. In addition, COBie testing and extraction tools are examined.

CIC BIM protocol

http://www.bimtaskgroup.org/bim-protocol/

The CIC BIM Protocol is a supplementary contract agreement for appointments by Construction Clients and Contractor Clients. It covers BIM model production and delivery requirements and also sets out information requirements. The protocol can be included in a contract or appointment by a simple amendment.

Employers information requirements

http://www.bimtaskgroup.org/bim-eirs/

The Employer's Information Requirements (EIRs) are included in tender and appointment documents, defining model requirements and outputs at each stage. The EIRs cover Technical, Management and Commercial aspects of the requirements and are detailed on the website.

Scope of services for information management

http://www.bimtaskgroup.org/scope-of-services-for-information-management/

These documents detail the Information Management role that is fundamental to BIM delivery on a project, managing the Common Data Environment, project information and facilitating collaborative working, information exchange and project team management. The role does not involve design responsibility. However, it could be carried out by a consultant with design responsibility, or the Main Contractor.

Pas1192/Part2

http://www.bimtaskgroup.org/pas11922-overview/

'The purpose of the Publicly Available Specification (PAS) is to support the objective to achieve BIM maturity Level 2 by specifying requirements for this level, setting set out the framework for collaborative working on BIM enabled projects and providing specific guidance for the information management requirements associated with projects delivered using BIM.' – Task Group Website

The PAS is the key overarching document that builds upon BS 1192:2007, defining the BIM processes for the Common Data Environment on a project for delivery from the start at definition of need through to handover, and detailing required management processes in a multi-disciplinary BIM environment. PAS 1192/Part 3 to be developed later this year will detail information management for an operational asset to support maintenance and portfolio management activities.

Insurance guidance note

http://www.bimtaskgroup.org/professional-service-indemnity-insurance-guidance/

The CIC has carried out extensive consultation with the Professional Indemnity insurance market, and developed some simple guidance for all those involved in design in a BIM environment. Guidance documents are provided on the website.

Video resources (under resources tab)

http://www.bimtaskgroup.org/video-resources/

Various leading members of the Task Group have produced videos that give an overview of the Government BIM programme covering aspects such as Education and Training, Commercial, Technical and Government Soft Landings.

Conclusion

This is a defining moment in the Government BIM initiative.

At a time when the UK has recently moved into second place in the world in terms of BIM adoption, outranked only by Finland, this comprehensive issue of documents provides much needed guidance and standards for all stakeholders involved in implementing and using BIM on their projects. As these documents become embedded in industry practice, they will provide a platform to move forward in BIM adoption and could well prove to be a landmark in the transformation of our industry.

RIBA plan of work 2013

While not a UK BIM Taskgroup document, the RIBA Plan of Work is an industry standard, and has been for over 50 years, setting out project processes. The update to be issued in May 2013 moves the RIBA PoW into the digital environment, available as a free online customisable tool. It aligns with the overall Digital Plan of Work.

Go to www.ribaplanofwork.com

Growth through BIM

Richard Saxon, CIC UK BIM Ambassador, has produced a report 'Growth through BIM' which documents BIM in the UK, potential developments and the implications for the next few years. It is an excellent piece of work. As a 'state of the nation' report, this is a go-to document and compulsory reading for anyone interested in BIM in the construction industry.

Growth through BIM is available as a free download at:

http://www.cic.org.uk/news/article.php?s=2013-04-25-cic-publish-growth-through-bim-by-richard-g-saxon-cbe

Key web resources

BIM Task Group – http://www.bimtaskgroup.org

Construction Industry Council – http://www.cic.org.uk

BIM Regional Hubs – http://www.bimtaskgroup.org/cic-bim-regional-hubs/

Please note that all references to the websites are accurate at the time of publication.

Briefing Note 3.06   Project planning

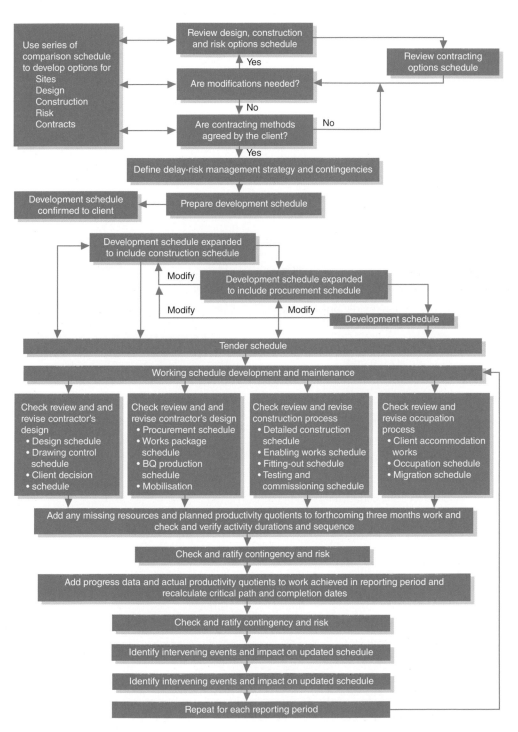

Project planning.

Sl. No.	Characteristic	Traditional	Design and build	Management contracting	Construction management
1.	Diversity of responsibility	Moderate	Limited	Large	Large
2.	Size of market from which costs can be tested	Moderate	Limited	Moderate	Large
3.	Timing of predicted cost certainty	Moderate	Early	Late	Late
4.	Need for early precise definition of client requirements	Yes	Yes	No	No
5.	Availability of independent assistance in development of design brief	Yes	No	Yes	Yes
6.	Speed of mobilisation	Slow	Fast	Fast	Fast
7.	Flexibility in implementing changes	Reasonable	Limited	Reasonable	Good
8.	Availability of recognised standard documentation	Yes	Yes	Yes	Limited
9.	Ability to develop proposals progressively with limited and				
10.	progressive commitment	Reasonable	Limited	Reasonable	Good
	Cost-monitoring provision	Good	Poor	Reasonable	Good
	Construction expertise input to design	Moderate	Good	Good	Good
11.	Management of design production				
12.	programme	Poor	Good	Good	Good
13.	Client influence on trade contractors	Limited	None	Good	Good
14.	Provision for controlling quality of construction materials and workmanship	Moderate	Moderate	Moderate	Good
15.	Opportunity for contractor to exploit cash flow	Yes	Yes	Yes	No
16.	Financial incentive for contractor to manage effectively	Strong	Strong	Weak	Minimal
17.	Propensity for confrontation	High	Moderate	Moderate	Minimal

The significance of the features listed in the table above may be outlined as follows:

1. Having widely dispersed responsibilities for different activities may provide the project manager with greater control, for example, in the selection of preferred consultants. It may, however, make it difficult to pinpoint responsibility.

2. It is acceptable practice to limit the numbers of tenders invited from contractors based on value and design development criteria. Where tendering involves significant design development, the cost will discourage contractors unless invitations are restricted. Such restrictions may not produce the most competitive price available unless careful pre-tender assessments are made.

3. Although the establishment of a certain financial outcome at an early stage in the development programme will minimise client risks, it could well be at a price. This is because of the risks which the tenderers will have to assume. A balance has to be achieved that depends on all the circumstances.

4. The client's requirements document associated with design and build procurement is a definitive statement. It must be produced early and it becomes the basis for all subsequent activities. Other procurement options enable progressive development of the client's brief, which may be helpful where there is uncertainty or greater complexity.

5. Independent assistance with the development of a design brief, which is integral to procurement options, may be advantageous where there is uncertainty or greater complexity, similar to item 4.

6. Mobilisation of construction using traditional procurement is relatively slow because much of the design development must be completed before appointment of the contractor, whereas all other methods enable progressive design and construction.

7. Little flexibility to accommodate variations exists within the design and build method. The other methods make reasonable provision for flexibility through the issuing of variations or additional works contracts.

8. Standard documentation, with which the industry is familiar, allows agreements to be entered readily. Although it enables incorporation of particular requirements, the drafting of unique documents often involves much negotiation and expense.

9. Where there are significant uncertainties or where limited finance is available, the opportunity to develop and appraise proposals may be advantageous. There may even be an opportunity to carry out construction on a progressive basis, step by step.

10. All procurement methods should seek to provide facilities for client cost monitoring, although the possible detail will vary.

11. Contractors' input to design could produce more cost-effective solutions provided contractors' interests are accommodated correctly. By using design and build, the contractor clearly has a vested interest in providing such input.

12. The schedule of preparation of production information is often critical to, and should be determined by, the construction schedule.

13. Procurement methods have different abilities to select preferred trade or works contractors that actually execute the works; no influence in selection is possible using design and build, and only limited influence is possible using the traditional method.

14. Design and build procurement makes no provision for the monitoring of construction quality – any monitoring required by the client must be independently commissioned. In other forms of procurement members of the design team – or the management contractor or the construction manager – may have monitoring responsibilities. But in all cases, except the last, only limited control of quality is available.

3.07 Briefing Note

15. Design and build standard forms usually make no provision for the provision of a working schedule. Accordingly, it is essential to effective project control that the necessary amendments are made to the standard form to facilitate the control of time to enable the project manager to properly execute his responsibilities in that regard.

16. Since construction works involve substantial financial transactions, there is considerable financial benefit to the main contractor in achieving payments as promptly as possible while delaying payments due as long as possible. This may have a significant detrimental effect on the attitudes and performance of the specialist subcontractors, and hence on the quality of their workmanship, thus exacerbating the limited quality control characteristics of procurement methods. Where payments to specialist contractors are under direct control of the client/construction manager, this can be turned to advantage.

17. Management procurement methods provide for remuneration of the management contractor or construction manager on the basis of a fee, not necessarily related to performance. Equitable performance measurement is often difficult. In design and build procurement, there is a strong incentive for good management; in traditional procurement, there is also an incentive for the contractor.

18. Construction quality, speed and cost can all be improved through good teamwork. Procurement methods which recognise the varying responsibilities of those managing construction operations and which preclude exploitation of any party are most likely to avoid confrontation.

Selection of the procurement method

From the foregoing, it can be seen that the most important characteristics of each method will best suit particular types of project. For example, design and build procurement would be an obvious choice where a client has limited interest in involvement with the design or construction process; and when there are clearly defined, straightforward requirements, including a need for early determination of cost.

It is necessary to consider all the characteristics of the project and to compare them with the characteristics of the various procurement methods available. The most important characteristics should be identified initially, after which secondary and peripheral issues should be considered, and the details determined for any necessary adaptation of the basic procurement methods available. For example, although design and build procurement may appear well suited to the project characteristics, it will probably be appropriate for the client to appoint an architect or planning consultant to progress the project through planning approval. The documentation produced would then be incorporated in the client's requirements on which design and build tenders would be sought.

Care must be taken in adapting any particular procurement method to compensate for perceived shortcomings, to avoid compromising the basic principles and essential characteristics. Thus, for example, although the engagement of design assistance for preparation of the client's requirements will inevitably dilute the single-point responsibility attribute of design and build procurement, the effects of this dilution should be mitigated by careful definition of responsibilities and terms of engagement. Similar care must be exercised when procuring specialist components or services incorporating design elements within a traditional project procurement method.

Selection of a procurement method is thus an essential element in the development of the policies to be adopted for implementation of all projects. In view of the fundamental differences in philosophy between the four basic procurement methods, the method should be determined at the earliest possible stage so that timely decisions can be made on the engagement of appropriate project resources. The development process can be optimised only by giving consideration at the earliest stage to the issues on which the appropriate procurement method should be determined.

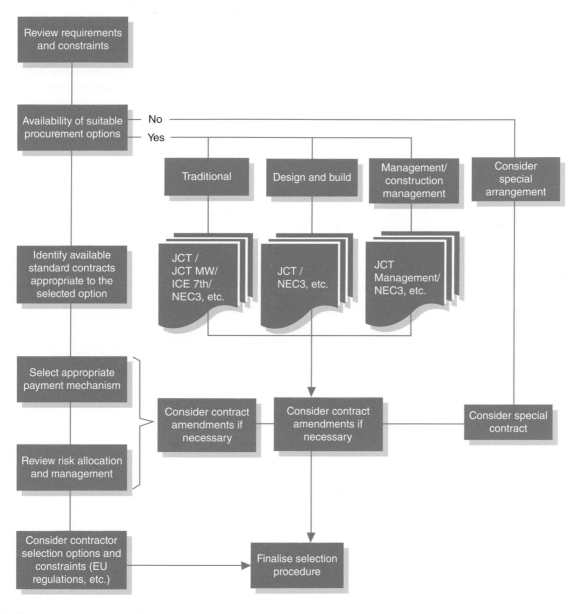

Selecting a procurement route.

Briefing Note 3.08   Framework agreements

Framework agreements can be described as agreements to provide both goods and services on predefined and specified terms and conditions with a selected number of suppliers (e.g. consultants, designers and constructors). While entering into a framework agreement does not by implication necessitate a binding requirement on either side to procure or provide the goods or services, the framework agreement will usually specify the terms that will apply if and when they are provided. There are a wide variety of framework procurement contracts including forms published by JCT, NEC and OGC.

The current EU Directives (2004/18/EC1 – Article 32) and the UK Regulations (enacted 31 January 2006 Regulation 19) also expressly provide for this form of procurement. It is to be noted that a framework agreement is different from a framework contract.

What are the advertising requirements?

The OJEU contact notice, as a minimum, must:

- make it clear that a framework agreement is being awarded

- include the identities of all the contracting authorities entitled to call-off under the terms of the framework agreement

- state the length of the framework agreement; usually it will be a maximum of four years

- include the estimated total value of the goods, works or services for which call-offs are to be placed and, so far as is possible, the value and frequency of the call-offs to be awarded

How is the framework agreement awarded?

- Open, restricted or, in certain circumstances, negotiated or competitive dialogue processes can be used.

- Award can be made to one or multiple providers (at least three).

- Mandatory standstill period applies – only for the agreement – not for the future call-offs.

What is the process for a call-off?

- When awarding individual contracts under framework agreements (called call-offs), authorities do not have to go through the full procedural steps in the EU Directives again.

- The weighting criteria can be varied subject to certain conditions, that is, the criteria used for mini-competition can be different from those used for the framework award.

- Call-offs may not need mini-competitions if the terms laid down in the framework agreements are sufficiently precise. However, how this is to be done is not set out in the regulations.

- Not all suppliers need to be included in the mini-competitions, particularly if the framework is split in different 'lots'.

The process charts for framework agreements and call-offs are set out later.

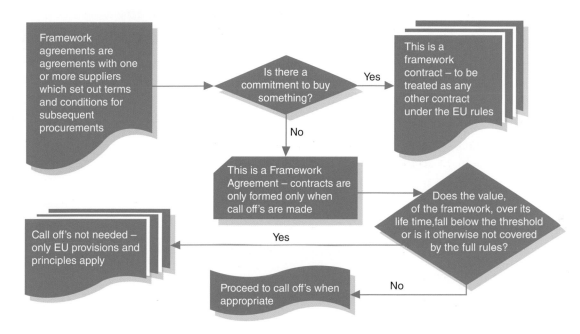

Framework agreements.

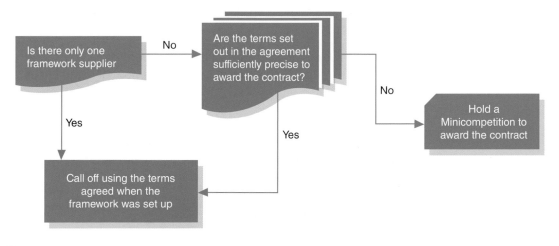

Call-off stage.

Briefing Note 3.09　Procedure for the selection and appointment of consultants

Stages	Key steps
Strategy	■ Decide works procurement strategy ■ Prepare project brief ■ Prepare consultant's brief ■ Decide terms of engagement including the choice of single/multiple appointment and phased appointment
Pre-selection	■ Prepare preliminary list ■ Decide criteria of selection
Selection	■ Invite to tender ■ Evaluate tender ■ Assess tender
Appointment	■ Finalise terms of engagement ■ Finalise management, monitoring and review process

Guidance for selection process

1. Determine what duties are to be assumed by the consultants and prepare a schedule of responsibilities. If applicable, consider what level of in-house expertise is available.

2. Check to see if the client has any in-house procedures or standard conditions of engagement for the appointment of consultants and what scope there is for deviating from them.

3. Decide on the qualities most needed for the project, and the method of appointment. Agree them with the client.

4. Establish criteria for evaluating consultants with weighting values (e.g. 5 vital, 0 unimportant) for each criterion.

5. Assemble a list of candidates from references and recommendations. Check any in-house approved and updated lists of consultants.

6. Prepare a shortlist by gathering information about possible candidates. Check which firms or individuals are prepared in principle to submit a proposal.

7. Assess candidates against general criteria and invite proposals from a select number (no more than six or less than three per discipline). Invitation documents should be prepared in accordance with the checklist given later. Competitive fee bids, if required, should conform to relevant codes of practice.

8. Arrange for conditions of engagement to be drawn up. The conditions, the form of which will vary with the work required and the type of client, should refer to a schedule of responsibilities for the stages for which the consultant is appointed and include a clause dictating compliance with the project handbook. The conditions of engagement should be based as closely as possible on industry standards (e.g. as set by CIC, RIBA, ACE, NEC and RICS). Consistency of style and structure between conditions for different members of the team will improve each member's understanding of their own and others' responsibilities. Each set should include this aspect. Fee calculation and payment terms should be clearly defined at the outset, together with the treatment of expenses, that is, included or not in the agreed fee.

9. Determine the criteria for assessing the consultants' proposals. Agree them with the client.

10. Appraise the proposals and select the candidates most appropriate for the project. Proposals should be analysed against the agreed criteria using weighting analysis.

11. Arrange final interview with selected candidates (minimum of two) for final selection/negotiation as necessary.

12. Submit a report and recommendation to the client.

13. Client appoints selected consultant.

14. Unsuccessful candidates are notified that an appointment has been made.

Checklist

1. The Consultant's brief must include:

    - project objectives

    - requirements of other participants

    - services to be provided

    - project schedule including the key dates

    - requirement of reports including key dates

2. Invitation documents must include:

    - a schedule of responsibilities

    - the form of interview panel

    - draft conditions of engagement (an indication of the type to be used)

    - design skills or expertise required

    - personnel who will work on the project, their roles, time-scales, commitment, output

    - warranties required, for whose benefit and in what terms

Invitations should ask candidates to include information on the level of current professional indemnity insurance cover for the duration of the project. Details of policy, date of expiry and extent of cover for subcontracted services must be provided.

3.09 Briefing Note

Example of consultancy services at different project stages

The following is a brief list of consultancy services typically provided at different stages of a project. However, this neither is a comprehensive list nor outlines the preferred sequence as that may vary from project to project. For a detailed scope of services, documents such as the CIC Scope of Services can be consulted.

Inception/feasibility

- Identification of client requirements, objectives and a commitment to sustainability including preparation of the project brief

- Feasibility studies including evaluation of options, environmental impact assessment, site assessment, planning guidance and commercial assessment

Strategy/pre-construction

- Design development including preparation of outline design and scheme design

- Development of cost estimates, tender preparation and evaluation and preparing project schedule

- Preparation of construction specifications and schedules

Construction/commissioning

- Preparation and issuing working drawings and variations

- Project/construction management

- Inspection, monitoring and valuation of construction

- Certification of payment

- Advising on dispute resolution

- Confirmation of completion

- Assisting in project handover

Completion/handover

- Ensuring defects correction

- Settlement of project including final accounts

- Confirmation of operation and maintenance procedures

- Post-project appraisal and feedback

Briefing Note 3.10   Selection and appointment of contractors

Pre-tender process

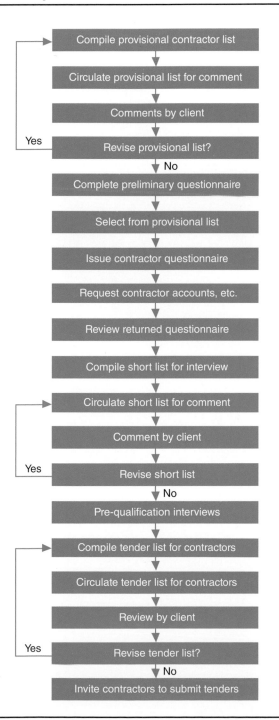

Compile provisional contractor list

↓

Circulate provisional list for comment

↓

Comments by client

↓

Revise provisional list?

Yes ← / ↓ No

Complete preliminary questionnaire

↓

Select from provisional list

↓

Issue contractor questionnaire

↓

Request contractor accounts, etc.

↓

Review returned questionnaire

↓

Compile short list for interview

↓

Circulate short list for comment

↓

Comment by client

↓

Revise short list

Yes ← / ↓ No

Pre-qualification interviews

↓

Compile tender list for contractors

↓

Circulate tender list for contractors

↓

Review by client

↓

Revise tender list?

Yes ← / ↓ No

Invite contractors to submit tenders

Checked by   Comments

Initial questionnaire

	Form Ref:
Ref. number:	
Contract title:	

Item no.	Question	Response
1.0	Turnover of company?	
2.0	What is the value of contracts secured to date?	
3.0	What is the largest current contract?	
4.0	Is the contractor willing to submit a tender?	
5.0	Is the contractor willing to work with all team members?	
6.0	Is the contract period acceptable?	
7.0	If not how long to complete works?	
8.0	Is the anticipated tender period acceptable?	
9.0	If not how long to tender?	
10.0	What are the mobilisation periods of (a) completion of drawings? (b) fabrication? (c) start on site from order?	
11.0	Is the labour used direct self-employed or subcontract?	
12.0	What element of the contract will be sublet?	
Comments		
Signature and date _____		

Selection questionnaire

1 Name of company:

2 Address:

3 Telephone no.:   Facsimile no.:

4 Nature of business:

5 Indicate whether

(a) manufacturer   ☐

(b) supplier   ☐

(c) subcontractor   ☐

(d) main contractor   ☐

(e) design and build contractor   ☐

(f) management contractor   ☐

6 Indicate whether   ☐

(a) sole trader   ☐

(b) partnership   ☐

(c) private limited   ☐

(d) public limited   ☐

7 Company registration number:

8 Year of registration:

9 Bank and branch:

10 VAT registration number:

11 Tax exemption certificate number: Date of expiry:

12 State annual turnover of current and previous four years:

13 State value of future secured work:

14 State maximum and minimum value of works undertaken:

15 Are you registered with any QA accreditator?

16 State previous projects undertaken with this company

17 Are you prepared to sign a design warranty?

18 Are you prepared to provide a performance bond?

19 Are you prepared to provide a parent company guarantee?

20 State details of your PL, EL and PI insurances

21 State your environmental accreditations

22 State your health and safety compliance accreditations

23 Do you have a safety policy?

24 Are you willing to act as principal contractor under CDM?

   Provide evidence of competency as a principal contractor as defined under CDM

25 Which elements do you sublet?

26 List of similar projects

   Project 1.:

   Address:

   Architect:

   Contact: Telephone no.:

   Contractor:

   Contact: Telephone no.:

   Value:

   Year completed:

   Project 2.:

   Address:

   Architect:

   Contact: Telephone no.:

   Contractor:

   Contact: Telephone no.:

   Value:

   Year completed:

   Project 3.:

   Address:

   Architect:

   Contact: Telephone no.:

   Contractor:

   Contact: Telephone no.:

   Value:

   Year completed:

3.10 Briefing Note

Pre-qualification interview agenda

Ref. number:

Contract title:

1.0	**Introduction**
1.1	Purpose of meeting
1.2	Introduction to those present
2.0	**Description of overall project and schedule**
2.1	General description of the project
2.2	Master schedule in summary
2.3	General description of contract
3.0	**Explanation of contract terms and conditions**
3.1	Outline and scope of contract
3.2	Responsibilities of the contractor
3.3	Outline of contract conditions including any significant amendments
3.4	Schedule
3.5	Specification
3.6	Drawings
3.7	Preliminaries
3.8	Budget prices
4.0	**Project organisation**
4.1	Site administration and project team
4.2	Setting out and dimensional control
4.3	Materials handling and control
4.4	Site establishment
4.5	Contractor supervision and on-site representative
4.6	Labour relations
4.7	Quality management
4.8	Health and safety plan
5.0	**Tendering**
5.1	Period of tendering
5.2	Mid-tender interview
5.3	Tender return date address and contact name
6.0	**Actions required**
6.1	Summary of actions and date deadlines

3.10 Briefing Note

Tendering process checklist

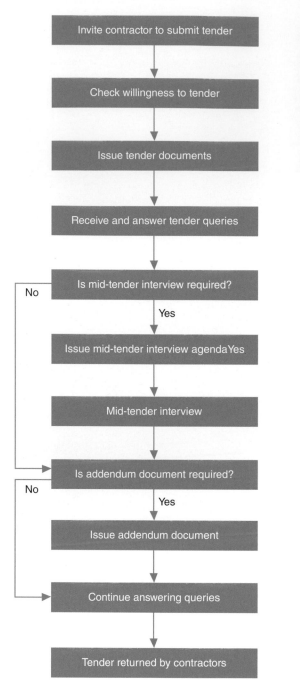

Checked by	Comments

Invite contractor to submit tender

Check willingness to tender

Issue tender documents

Receive and answer tender queries

Is mid-tender interview required? — No

Yes

Issue mid-tender interview agendaYes

Mid-tender interview

Is addendum document required? — No

Yes

Issue addendum document

Continue answering queries

Tender returned by contractors

Tender document checklist

Ref. number:

Contract title:

- ☐ Invitation to tender
- ☐ Introduction and scope of contract
- ☐ Instructions to tenderers
- ☐ Form of tender
- ☐ General preliminaries
- ☐ Particular preliminaries
- ☐ Form of contract and amendments
- ☐ Contract schedule
- ☐ Method statement
- ☐ Quality management
- ☐ Project health and safety plan
- ☐ Project resource details
- ☐ Specification
- ☐ List of drawings
- ☐ Bill of quantities or pricing schedule
- ☐ General summary
- ☐ Declaration of non-collusion
- ☐ Performance bond
- ☐ Warranty
- ☐ Preconstruction information
- ☐ Survey reports (soil, contamination, etc.)

Other documents (please list below)
- ☐
- ☐
- ☐
- ☐
- ☐
- ☐
- ☐
- ☐

Mid-tender interview agenda

Ref. number:

Contract title:

1.0 Introduction

1.1 Purpose of meeting

1.2 Introduction of those present

2.0 Confirmation of further information issued

3.0 Responses to existing queries

3.1 Contractor

3.2 Architect

3.3 Civil and structural engineer

3.4 Mechanical and electrical engineer

3.5 Other consultants

3.6 Quantity surveyors

3.7 Project manager

4.0 Other additional information

4.1 Contractor

4.2 Architect

4.3 Civil and structural engineer

4.4 Mechanical and electrical engineer

4.5 Other consultants

4.6 Quantity surveyors

4.7 Project manager

5.0 Contractor's queries

6.0 Confirmation of tender arrangements

6.1 Date

6.2 Time

6.3 Address

7.0 Any other business

Returned tender review process

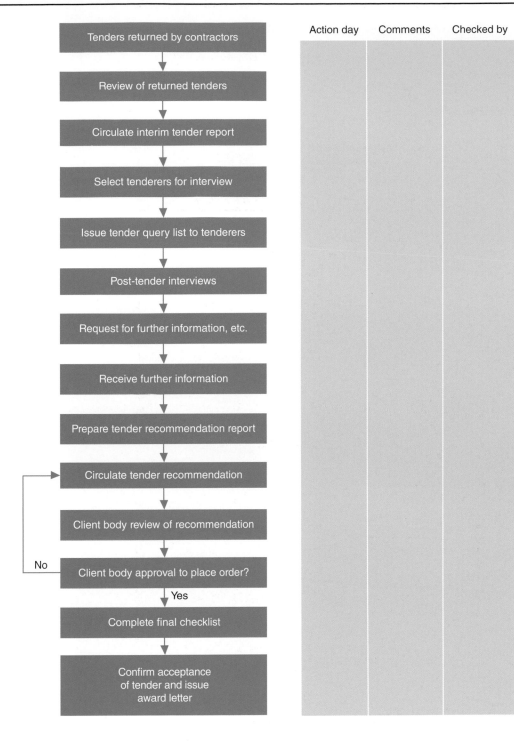

	Action day	Comments	Checked by
Tenders returned by contractors			
Review of returned tenders			
Circulate interim tender report			
Select tenderers for interview			
Issue tender query list to tenderers			
Post-tender interviews			
Request for further information, etc.			
Receive further information			
Prepare tender recommendation report			
Circulate tender recommendation			
Client body review of recommendation			
Client body approval to place order?			
Complete final checklist			
Confirm acceptance of tender and issue award letter			

No

Yes

3.10 Briefing Note

Returned tender bids record sheet

Ref. number:			
Contract title:			
Allocated budget: £		**Programme period:**	
No.	**Contractor qualifications, etc.**	**Prog.**	**Bid sum**
1.			
2.			
3.			
4.			
5.			
6.			
7.			
8.			

Signed by the undersigned as a true record of duly and properly

Received tender bids for _____

On this day _____ (day) _____ (month) _____ in the year _____

Signed:

Company:

Signed:

Company:

Signed:

Company:

Signed:

Company:

Post-tender interview agenda

Ref. number:

Contract title:

1.0   Introduction

1.1   Introduction to those present

1.2   Purpose of meeting

2.0   Confirmation of contract scope and responsibilities

3.0   Detailed bid discussions

3.1   Contractual

3.2   Cost

3.3   Schedule

3.4   Method

3.5   Technical matters

3.6   Resourcing matters

3.7   Supplier/subcontractor matters

3.8   Health and safety matters

3.9   Quality and environmental management

4.0   Contractor queries

5.0   Action and responses

5.1   Agreement of action items

5.2   Agreement of deadline dates for resolution of action items

3.10 Briefing Note

Final tender evaluation report

Ref. number:

Contract title:

1.0 Summary of final tender bids following post-tender interviews

2.0 Cost appraisal

3.0 Schedule appraisal

4.0 Method statement appraisal

5.0 Technical appraisal

6.0 Contractual appraisal

7.0 Quality and environmental management appraisal

8.0 Health and safety appraisal

9.0 Resourcing appraisal

10.0 Recommendation to place contract

Appendices

1 Completed tender bid of recommended tenderer

2 Addenda and other information issues during tendering

3 Mid-tender interview meeting minutes

4 Query lists and responses

5 Post-tender meeting minutes

6 Any other letters, etc., since tender issue

7 Summary of procurement status

3.10 Briefing Note

Approval to place contract order

Ref. number:

Contract title:

In accordance with clause of the _____

(Contract Form) _____

We _____

Do not have any objection to the placing of a contract with _____

for _____

All in accordance with the tender recommendation report submitted to us on the _____ (date).

In submitting the tender recommendation report the consultants are fully satisfied that the contractor has complied in full with the tender documents.

Signed by: _____

Signed by: _____

Signed by: _____

Final general checklist

Ref. number:

Contract title:

Check once again that the following were carried out:

☐ Long list

☐ Telephone selection questionnaires

☐ Contractor selection questionnaires, company accounts, references and any reports of visits to offices, factories and contracts

☐ Short list

☐ Pre-qualification interview minutes

☐ Tender list

☐ Substitute tender list

☐ Tender documents and checklist

☐ Tender query lists addendum letters prior to mid-tender interviews

☐ Mid-tender interview minutes

☐ Tender query lists addendum letters, etc., post mid-tender interviews

☐ Returned tender summary form and returned tender documents

☐ Interim tender analysis and recommendations report

☐ Post-tender query lists to contractors

☐ Post-tender interview minutes

☐ Post-tender addendum letters, etc.

☐ Final tender analysis and recommendations report

☐ Contractor acceptability final check

☐ Approval to place contract order

3.10 Briefing Note

Briefing Note 3.11   Guidance on EU procurement rules

The European Union (EU) Procurement Directives, and the regulations that implement them in the UK, set out the law on public procurement. Their purpose is to open up the public procurement market and to ensure the free movement of goods and services within the EU.

The rules apply to purchases by public bodies and certain utilities which are above set monetary thresholds. They cover all EU Member States and, because of international agreements, their benefits extend to a number of other countries.

Where the regulations apply, contracts must be advertised in the *Official Journal of the European Union* (OJEU) (unless it qualifies for a specific exclusion, e.g. on grounds of national security) and there are other detailed rules that must be followed. The rules are enforced through Member States' courts, and the European Court of Justice (ECJ).

What are the key changes?

The changes introduced through the current set of regulations (enacted on 31 January 2006) include:

- supply, services and works are consolidated into a single set of regulations
- framework agreements and e-auctions expressly included
- a new competitive dialogue procedure introduced in addition to the open and restricted procedures
- a dynamic purchasing system introduced
- specific provisions made for central purchasing bodies
- mandatory exclusion of entities whose directors or other decision-makers have been convicted of certain offences
- a 10-day standstill period at the award stage prior to contract signature has been provided for

What about mixed contracts?

- Where a contract covers both services and supplies, the classification should be determined by the respective values of the two elements.
- Where a contract covers works/supplies or works/services, it should be classified according to its predominant purpose.
- Where a contract provides for the supply of equipment and an operator, it should be regarded as a services contract.
- Contracts for software are considered to be for supplies unless they have to be tailored to the purchaser's specification, in which case they are services.

What is the advertisement requirement?

Generally, contracts covered by the regulations must be the subject of a call for competition by publishing a contract notice in the OJEU. In most cases, the time allowed for responses or tenders must be no less than a set period, although some reduction is possible under certain circumstances (see SIMAP website for further details).

There are some services (categorised as Part A and Part B services) where a reduced advertisement requirement applies: details of this are available on SIMAP website (http://simap.europa.eu).

The following table outlines the advertisement timescale requirements.

Procedure	Text	Days*
Open	Minimum time for receipt of tenders from date contract notice sent.	52
	Reduced when prior information notice (PIN) published (subject to restrictions) to, generally, 36 days and no less than 22 days	
Restricted	Minimum time for receipt of requests to participate from the date contract notice sent	37
	Minimum time for receipt of tenders from the date invitation sent.	40
	Reduced when PIN published (subject to restrictions) to, generally, 36 days and no less than 22 days	
Restricted accelerated	Minimum time for receipt of requests to participate from the date contract notice sent	15
	Minimum time for receipt of tenders from the date invitation sent	10
Competitive dialogue and competitive negotiated	Minimum time for receipt of requests to participate from the date contract notice sent	37
Competitive negotiated accelerated	Minimum time for receipt of requests to participate from the date contract notice sent	15

*Timescales are correct at the time of the preparation of the document. Dispensations are available if electronic means of communications are used

What are the procurement options?

- Open procedure: All interested parties can respond.

- Restricted procedure: A selected number of respondents are invited to tender.

- Competitive dialogue procedure: Following an OJEU contract notice and a selection process, the authority then enters into dialogue with potential bidders, to develop one or more suitable solutions for its requirements and on which chosen bidders will be invited to tender.

- Negotiated procedure: A purchaser may select one or more potential bidders with whom to negotiate the terms of the contract. An advertisement in the OJEU is usually required but, in certain circumstances, described in the regulations, the contract does not have to be advertised in the OJEU. An example is when, for technical or artistic reasons or because of the protection of exclusive rights, the contract can only be carried out by a particular bidder.

What is the impact of the regulations on private sector projects?

For public works concession contracts (i.e. contracts under which the contractor is given the right to exploit the works, e.g. tolled river crossings), the winning concessionaire is required to comply with certain OJEU advertising requirements for works contracts which it intends to award to third parties. For some subsidised works contracts (civil engineering activities, building work for hospitals, facilities intended for sports, recreation and leisure, school and university building or buildings for administrative purposes), the public authority awarding the grant is obliged to require the subsidised body to comply with the regulations, as if it were a public authority, as a condition of grant. This provision has, for example, been invoked for many lottery-funded projects. There is a similar requirement for subsidised service contracts in connection with subsidised works.

Endnote

This guidance is not intended as a substitute for project-specific legal advice, which should always be sought by a public authority where required. The EU procurement regime is not static. It is subject to change, driven by evolving European and domestic case law, European Commission communications, new and revised Directives and amendments of the existing UK regulations. Further information can be obtained from SIMAP or the Government procurement authorities.

3.11 Briefing Note

Governance principles

1. Good practice guidance suggests a number of principles that should underpin project governance.[4,5]

2. The board has overall responsibility for governance of project management.

3. The roles, responsibilities and performance criteria for the governance of project management are clearly defined.

4. Disciplined governance arrangements, supported by appropriate methods and controls, are applied throughout the project lifecycle.

5. A coherent and supportive relationship is demonstrated between the overall business strategy and the project portfolio.

6. All projects have an approved plan containing authorisation points at which the business case is reviewed and approved. Decisions made at authorisation points are recorded and communicated.

7. Members of delegated authorisation bodies have sufficient representation, competence, authority and resources to enable them to make appropriate decisions.

8. The project business case is supported by relevant and realistic information that provides a reliable basis for making authorisation decisions.

9. The board or its delegated agents decide when independent scrutiny of projects and project management systems is required, and implement such scrutiny accordingly.

10. There are clearly defined criteria for reporting project status and for the escalation of risks and issues to the levels required by the organisation. The organisation fosters a culture of improvement and of frank internal disclosure of project information.

11. Project stakeholders are engaged at a level that is commensurate with their importance to the organisation and in a manner that fosters trust.

The four areas of governance that will help organisations to deliver these 11 principles are:

- portfolio direction

- project sponsorship

---

4   Directing Change a guide to governance of project management (2005). www.apm.org.uk (accessed April 2014).

5   Also see Morgan, A. & Gbedemah, S. (2010) How poor project governance causes delays. A paper presented to the *Society of Construction Law at Meeting in London*, 2 February 2010.

- project management
- disclosure and reporting

Portfolio direction

Portfolio direction is concerned with ensuring that the project portfolio (which, in context of construction and development projects is akin to the scheme) is aligned with the organisation's objectives including profitability, customer service, reputation and sustainability:

- Is the organisation's project portfolio aligned with its key business objectives, including those of profitability, customer service, reputation, sustainability and growth?

- Are the organisation's financial controls, financial planning and expenditure review processes applied to both individual projects and the portfolio as a whole?

- Is the project portfolio prioritised, refreshed, maintained and pruned in such a way that the mix of projects continues to support strategy and take account of external factors?

- Does the organisation discriminate correctly between activities that should be managed as projects and other activities that should be managed as non-project operations?

- Does the organisation assess and address the risks associated with the project portfolio, including the risk of corporate failure?

- Is the project portfolio consistent with the organisation's capacity?

- Does the organisation's engagement with project suppliers encourage a sustainable portfolio by ensuring their early involvement and by a shared understanding of the risks and rewards?

- Does the organisation's engagement with its customers encourage a sustainable portfolio?

- Does the organisation's engagement with the sources of finance for its projects encourage a sustainable portfolio?

- Has the organisation assured itself that the impact of implementing its project portfolio is acceptable to its ongoing operations?

Project sponsorship

Project sponsorship is the effective linkage between the senior management and the project management team. At its core is the corporate leadership and the decision making for the benefit of achieving the project objectives. It can be argued that the project sponsor role is the most pivotal for good project governance as this is the role that is most concerned with integration of the project objectives with the organisation's strategy. It is the communication route through which project managers report progress and issues upwards to the board and obtain authority and decisions on issues affecting their project. It takes ownership of the business case and is responsible for ensuring that the intended benefits become the project objectives and are delivered accordingly. Consequently, successful project sponsorship depends on the competence of the person or people employed to undertake these roles:

- Do all major projects have competent sponsors at all times?

- Do sponsors devote enough time to the project?

- Do project sponsors hold regular meetings with project managers and are they sufficiently aware of the project status?

- Do project sponsors provide clear and timely directions and decisions?

- Do project sponsors ensure that project managers have access to sufficient resources with the right skills to deliver projects?

- Are projects closed at the appropriate time?

- Is independent advice used for appraisal of projects?

- Are sponsors accountable for and do they own and maintain the business case?

- Are sponsors accountable for the realisation of benefits?

- Do sponsors adequately represent the project throughout the organisation?

- Are the interests of key project stakeholders, including suppliers, regulators and providers of finance, aligned with project success?

Project management

The effectiveness and efficiency of project management relates to the capacity, capability and competency of the project team to deliver the project objectives. Experience of good project management demonstrates that project risks that could lead to project failure are most effectively mitigated in organisations where there are strong competencies and effective management systems:

- Do all projects have clear critical success criteria and are they used to inform decision-making?

- Is the board assured that the organisation's project management processes and project management tools are appropriate for the projects that it sponsors?

- Is the board assured that the people responsible for project delivery, especially the project managers, are clearly mandated, sufficiently competent, and have the capacity to achieve satisfactory project outcomes?

- Are project managers encouraged to develop opportunities for improving project outcomes?

- Are key governance of project management roles and responsibilities clear and in place?

- Are service departments and suppliers able and willing to provide key resources tailored to the varying needs of different projects and to provide an efficient and responsive service?

- Are appropriate issue, change and risk management practices implemented in line with adopted policies?

- Is authority delegated to the right levels, balancing efficiency and control?

- Are project contingencies estimated and controlled in accordance with delegated powers?

Disclosure and reporting

This component of project governance is most reliant on the culture of the organisation. A culture of open and honest disclosure is paramount for effective reporting. Such a culture must flow from the project organisation throughout the supply chain.

3.12 Briefing Note

What is reported needs to be reliable and timely in order to enable the right decisions to be made at the right time for the project organisation. Without such timeliness the project is likely to fail:

- Does the board receive timely, relevant and reliable information of project forecasts, including those produced for the business case at project authorisation points?

- Does the board receive timely, relevant and reliable information of project progress?

- Does the board have sufficient information on significant project-related risks and their management?

- Are there threshold criteria that are used to escalate significant issues, risks and opportunities through the organisation to the board?

- Does the organisation use measures for both key success drivers and key success indicators?

- Is the organisation able to distinguish between project forecasts based on targets, commitments and expected outcomes?

- Does the board seek independent verification of reported project and portfolio information as appropriate?

- Does the board reflect the project portfolio status in communications with key stakeholders?

- Does the business culture encourage open and honest reporting?

- Where responsibility for disclosure and reporting is delegated or duplicated, does the board ensure that the quality of information that it receives is not compromised?

- Is a policy supportive of whistleblowers effective in the management of projects?

- Do project processes reduce reporting requirements to the minimum necessary?

Governance of project management is not the rigid application of a complex methodology. The best results will come from the intelligent application of principles combined with proportionate delegation of responsibility and the monitoring of internal control systems.

Briefing Note 3.13   Change management

Change in a construction project is any incident, event, decision or anything else that affects

- the scope, objectives, requirements or brief of the project

- the value (including project cost and whole-life cost) of the project

- the time milestones (including design, construction, occupation)

- risk allocation and mitigation

- working of the project team (internally or externally)

- any project process at any project phase

Changes during the design development process

The procedure outlined is used to control the development of the project design from the design brief to preparation of tender documents. It will include:

- addressing issues in the design brief

- variations from the design brief, including design team variations and client variations

- developing details consistent with the design brief

- approving key design development stages, namely, scheme design approval and detailed design approval

The procedure is based on the design development control sheet. The approved design will comprise the design brief and the full set of approved design development control sheets. The procedure comprises the following stages:

- The appropriate member of the design team addressing each design issue in the development of the brief, coordinated by the design team leader.

- Proposals developed are discussed with the appropriate members of the project's core group through submission of detailed reports/meetings coordinated by the project manager. Reports should not repeat the design brief, but expand it, address an issue and prepare a change.

- The design team leader coordinates preparation of a design development control sheet, giving:
  - design brief section and page references
  - a statement of the issue
  - a statement of the options
  - the cost plan item, reference and current cost
  - the effect of the recommendation on the cost plan and the schedule

    ☐ a statement as to whether the recommendation requires transfer of client contingency (i.e. a client variation to the brief) and if so the amount to be transferred

- The design team report section of the control sheet is signed by:
    - ☐ the design team member responsible for recommendations
    - ☐ the quantity surveyor (for cost effect)
    - ☐ the design team leader (for coordination)

- The design team leader sends the design development control sheet to the project manager who obtains the client's approval signature and returns it to the design team leader.

- The quantity surveyor incorporates the effect of the approved recommendation into the cost plan.

- The project manager incorporates the effect of the approval recommendation into the master schedule.

Design development control sheet

Client name:

Project name:

Sheet no.

Design team report

Design brief section:

Issue: Pages:

Options considered: 1.

                              2.

                              3.

Recommendation:

Cost plan item:

Ref:

Current cost:

Effect of recommendation on costs/schedules: Increase/decrease

Application for transfer of client contingency: Yes/No   Amount:

Architect/services engineer/structural engineer: Date:

Quantity surveyor: Date:

Design team leader: Date:

Client approval

Design development/Client contingency transfer approved (delete as applicable)

Position _____ Signature _____ Date _____

Example of change management process

- Identification of requirement for change
- Evaluation of change
- Consideration of implications and impact including risks
- Preparation of change order
- Reviewing of change order: client's decision stage
- Implement change
- Feedback including causes of change

Change order request form

Project no.	Date:	No.	
Client:			
Project:		**Distribution:**	
Subject – definition of change:			WHAT
Identified by:			WHO
Reasons for change:			WHY
	Discretionary	Non-discretionary	
Cost implication:			
Time implication:			
Recommended action:	Project manager	.........Date	
Client decision required by:	Date:		
Forwarded to client:	Date:		
Client's decision:	Date		
Projected schedule and cost plan (budget) amended on	Project manager		

Change order register

Project:			Client:	Job Id:	File reference:		
Request Id:	Date	Initiated by:	Description of change	Client decision required by:	Client decision obtained on:	Client decision	Client decision Id no.:

Briefing Note 3.14   Strategic collaborative working

What is Partnering?

Partnering is a management approach used by two or more organisations to achieve specific business objectives by maximising the effectiveness of each participant's resources. It requires that the parties work together in an open and trusting relationship based on mutual objectives, an agreed decision-making process and an active search for continuous measurable improvements.

Partnering is the most efficient way of undertaking all kinds of construction work including new buildings and infrastructure, alterations, refurbishment and maintenance. The basic elements have been in practice for a long time, but the benefits have not been recognised by those involved. The elements are evident when those involved on a project have worked together before and are brought into the project at the very early stages. In this ideal situation everyone naturally works together as a team.

Partnering can be based on a single project, but the real benefits are realised when it is based on a long-term strategic commitment. Project-specific partnering is about partnering on individual projects. Strategic partnering is about long-term relationships between parties who are prepared to work together over extended periods of time. By building on the individual strengths of the separate businesses, a strategic partnering arrangement can deliver steadily improved performance over several years.

Definition of project partnering

Project partnering is a set of actions taken by the work teams that form a project team to help them cooperate in improving their joint performance. Specific actions are agreed by the project team, taking account of the project's key characteristics, and its own experience and normal performance. The choice of actions is guided by a structured discussion of mutual objectives, decision-making processes, performance improvements and feedback.

Project partnering involves initial costs and provides substantial benefits. It is not a fixed way of working; it develops as project teams cooperate in finding the most effective ways of achieving agreed objectives.

Definition of strategic collaborative working

Strategic collaborative working is a set of actions taken by a group of firms to help them cooperate in improving their joint performance over a series of projects.

The actions initially aim to agree an overall strategy, ensure the right firms are included, financial arrangements support partnering, firms' cultures, processes and systems are integrated, overall performance is benchmarked, project processes are continually improved and the whole strategic partnering arrangement is guided by feedback. Ultimately, the actions aim at establishing and continuously developing a long-term business based on an integrated construction cycle that links clients' use of constructed facilities with their development and production.

The more partnering is used and the benefits are experienced, the more likely it is that strategic long-term partnering will be used. Partnering is about the formation and development of a relationship or relationships which benefit construction projects. Partnering creates a variety of opportunities and concerns for the participants. These include early involvement of suppliers, selection of all parties by value, performance measurement and continuous improvement, common team processes and commercial arrangements that align risk and reward for all parties on both the demand and supply sides of the industry. The term 'collaborative working' is often used to help eliminate misconceptions about previous definitions of partnering. This appears to encourage people to invest in learning about new best practice methods and adopt cooperative behaviour.

Essential features of partnering

Partnering consists of three essential features: mutual objectives, an agreed decision-making process and an active search for continuous improvement. They form the partnering logo shown as follows:

Mutual objectives

The most fundamental requirement for partnering is to agree mutual objectives. The aim is to find objectives that firmly establish for everyone involved in the project that their own best interests will be served by concentrating on the overall success of the project. When people cooperate in adopting a 'win-win' attitude they increase the chances that they will produce enough for everyone to have everything they reasonably want.

Partnering accepts that firms look after their own interests. It requires a tough-minded recognition by clients that they will get what they need only if consultants, contractors and specialists have a realistic opportunity to do good work and make reasonable profits. It requires an equally tough-minded recognition by consultants, contractors

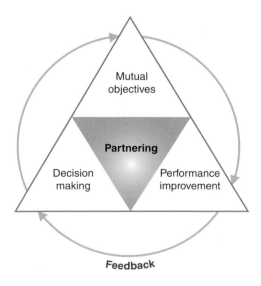

Essential actions of project partnering.

3.14 Briefing Note

and specialists that they prosper best when clients get excellent value, good buildings or infrastructure and no hassle. This focus on mutual objectives gives expression to the idea that when people cooperate they can produce more than enough to give everyone what they reasonably want. This is often described as a 'win-win' attitude in contrast to the traditional zero-sum assumption that if one person gains, someone else must lose.

Clients should ensure that agreed mutual objectives take account of the interests of everyone affected by the project. It may take time to deal with everyone's concerns. Inevitably the client, designers, managers, specialist contractors and manufacturers have different views about what constitutes success. And people tend to worry that in some way they will lose out if they cooperate in meeting other peoples' needs. Despite these perhaps natural reservations, experience shows that when project teams are brought together to discuss their individual interests they can find mutual objectives. The initial costs and time involved are worthwhile because well thought out mutual objectives avoid time and resources being wasted on designs likely to create problems at later stages.

Mutual objectives may deal with many process and product related issues but common subjects include:

- value for money

- guaranteed profits

- reliable quality

- fast construction

- handover to owner on time

- cost reduction

- costs within agreed budget

- operating and maintenance efficiency

- improved efficiency for users

- architectural quality

- a specific technical innovation

- excellent site facilities

- safe construction

- shared risks

- timely design information

- reliable flow of design information

- shared use of computer systems

- effective meetings

- training in decision-making skills

- training in management control systems

- no claims

In agreeing mutual objectives, it is essential to sort out the financial arrangements so that everyone gets a fair return in business terms. The worst situations in construction projects arise when any of the consultants or contractors is losing money.

Contractual arrangements should ensure that none of the firms, if they contribute their best efforts, would lose out relative to the others. The essential equity of good value for clients and fair profits for consultants, contractors and specialists provides the platform on which partnering flourishes.

Decision-making process

Construction projects bring together many work teams drawn from many different firms. These work teams need to agree how decisions will be made. The nature of decision-making systems is directly influenced by whether the client needs the project to produce an existing well-developed answer or an original design. An important consequence of this choice is the amount of time the client and his staff will need to spend in making decisions. Original designs take more time but should result in buildings and infrastructure that support the users' activities and delight everyone who sees them. Standard answers make fewer calls on the client's time, are quicker, cost less and provide reliable quality but may well impose more compromises on users and look dull.

The project team need to agree the information and communication systems they will use. They also need to decide the quality, time and cost-control systems they will use. They need to agree who will operate these systems and who will get the various outputs. They decide on the form and frequency of face-to-face meetings. They consider the use of taskforces, workshops, common project offices, social events and other ways of bringing teams closer together. Overall, the systems should ensure that good ideas are captured and properly considered. This means balancing the benefits of using an existing satisfactory answer, and the greater benefits, costs and time of finding and developing something better.

Whatever decision-making systems are agreed, they should include robust procedures to ensure that problems are resolved quickly in ways that encourage cooperative teamwork. This means most problems are resolved by the work teams directly involved. When a problem cannot be resolved in this way, it should be referred immediately to the project's core team and in exceptional cases to senior managers.

Continuous performance improvement

The main aim of partnering is to improve project teams' performance. Partnering that merely provides mutual objectives and agreed ways of making decisions will drift into inefficient ways of working. Partnering requires project teams to search for better answers. Project teams new to partnering should aim at one modest improvement that all members of the team regard as important. As experience of partnering grows, the scale and range of improvements will increase.

It is important that performance improvements in one area do not distract work teams from continuing to deliver their established normal performance in all other areas. This is an easy trap to fall into as attention is focused on the improvements and quality elsewhere slips without anyone noticing. This is why partnering procedures give explicit attention to the constraints of achieving normal performance as well as delivering performance improvements.

There is much discussion about the best ways of encouraging work teams to improve their performance. Many economists argue that competition provides the best incentives to improve performance. However, competition in the construction industry can easily become cut-throat and bid prices, quality and safety are driven down to levels that are hopelessly inefficient. The outcomes include claims, disputes, defects, late completions and good firms being driven out of business. Competition has a place in partnering to encourage consultants, contractors and specialists to invest in training and innovation to improve their own performance. This can be achieved even when there are long-term relationships between firms. By having two, three or four options available for key relationships, all the partners are motivated to continuously improve their performance.

There is usually provision in partnering contracts for incentives to be agreed, which may allow for participants to share in cost savings achieved, but may also provide for a similar distribution of losses due to errors or cost increases. These arrangements are known as gain share/pain share.

Benchmarking provides another weapon in the search for improved performance. Carefully researched information about best international practice is often used by experienced clients to guide the choice of targets. A good approach is to concentrate on whatever the client, consultants or contractors regard as their biggest problem.

There are advantages in project teams setting their own targets. When teams are given good information about the performance achieved by leading practice, they often set tougher targets than any they would accept from their managers.

Having agreed the performance improvements they will aim for, the best partnering teams try various ideas; continue with actions that work and change those that deliver no improvements. They often set up a task force to help find ways of meeting targets. This is a small group of people with relevant knowledge selected from within the project team and it may include external experts. Task forces should be given a short time to find an innovative answer that will deliver significant performance improvements.

The first partnering workshop should establish procedures to ensure that innovations and new actions found to deliver improvements will be built into standards and procedures for the benefit of the current and future projects.

Feedback

Teams need to be guided by feedback about their own performance if they are to deliver the substantial benefits that partnering can provide. Achieving performance improvements depends on project teams being provided with up-to-date and objectively measured feedback. Teams should measure their own performance and plot the results on control charts that show graphically how they are doing against their targets. Teams believe feedback they have produced themselves and use it to search for better ways of working. Feedback is most effective when it is expressed in positive terms. For example, quality should be measured by recording how often quality standards are achieved, not the number of failures.

Performance improves faster when successes are publicised and celebrated. It is vital that senior managers know when targets are being achieved and make a point of congratulating and rewarding the people involved. The rewards can be a token, but a dozen cans of lager presented at a light-hearted ceremony to the week's best work team can ensure that all the teams strive to be winners next week.

Failures must not be ignored. This is not to allocate blame which is counterproductive. Failures should be used to guide teams in looking for robust answers to problems so that performance is back on target quickly. Some effective teams make a point of celebrating failures because they provide opportunities to find more effective ways of working. When a failure arises, they have a party and then, with renewed enthusiasm, concentrate on finding a robust answer. It is important that senior managers are kept up to date about improvements in performance. This is essential if they are to remain committed to partnering. At least some managers in most organisations take a pride in being highly competitive and are sceptical of the idea that the cooperative methods used in partnering can possibly be effective. Without regular, well-founded feedback on the performance improvements delivered by partnering, there is always a risk that adversarial methods will be reintroduced.

Feedback should flow from project to project. Too many innovative ideas are lost because of weak feedback systems. Lessons need to be captured so that good ideas are applied on future projects and problems and defects do not recur. Leading firms involved in partnering have developed standards and procedures that systematically capture best practice as it emerges from their projects. The feedback-based standards and procedures help all their project teams concentrate on efficient work. This is an essential element in using strategic partnering and strategic collaborative working successfully.

Maintaining partnering throughout projects

The first partnering workshop on any project is important in giving the project team a firm basis for ensuring that partnering delivers benefits. However, best practice includes workshops throughout projects to review progress and if necessary change things agreed at the first partnering workshop. Change may be in response to the project going better than expected and the team realising they can aim for bigger performance improvements. It is perhaps more common for projects to face problems. These should be discussed at a workshop, which if the problem is sufficiently serious, should be called especially. The workshop should look for and agree actions that deal with persistent problems once and for all. Partnering is action oriented and dealing with problems quickly is central to its success.

A final workshop is used to identify good ideas and lessons identified during the project so they can be recorded and made available for use on future projects.

Partnering is an ongoing activity guided by workshops which all need to be taken seriously especially by the senior managers involved. The potential benefits are large and they are earned by concentrating on and continually reinforcing cooperative teamwork. CIOB's *Partnering in the Construction Industry: A Code of Practice for Strategic Collaborative Working* provides further detailed guidance and advice.

Briefing Note 3.15   PPP/PFI arrangements

The legal aspects relating to the PPP projects, in context of UK, are perhaps best explained through three different facets:

1. Pre-1997 Government policies

2. Post-1997 Government policies and

3. The European Commission guidance on PPP

However, it is necessary to point out that apart from the EC guidance, there is no specific Act or legislation controlling PPP projects in UK public sector – all the different sectors have created their own terms of engagement in terms of PPP projects, with input and support from regulatory, legislative and executive authorities. Although the first standard PFI contract was published in 1999, the different sectors have developed their own forms of contractual arrangements to suit their particular requirements.

In many sectors, there are non-statutory guidance which provides model documentation and advice in relation to the PPP processes. Some of these documentations are listed in Table 3.1.

Following the enactment of the Public Contracts Regulations (2006 No. 5), the use of the competitive dialogue procedure for PPP/PFI procurement is advised by the Government, with the negotiated procedure only to be used in exceptional circumstances. It is to be noted that the competitive dialogue procedure does not exist under the Utilities Contracts Regulations.

Standard documentations for PPP/PFI arrangements

Standard guidance (HMT publications)	Standardisation of PFI contracts (version 4) Change protocol principles Drafting pack for updating contracts Value for money assessment guidance
Defence	The MoD project agreement
Education (for BSF projects)	School standard form PFI contracts
Housing	Housing (HRA) model contract
ICT (OGC publication)	ICT model contract and guidance
Waste management (Waste infrastructure delivery programmes)	WIDP procurement pack WIDP guidance documents
Local authorities	Fire & rescue & police service guidance. Change protocol.

	Social care guidance.
	Streetlighting procurement pack and model contract.
	Joint services guidance.
	Standardisation of waste management PFI contracts
Operational task force	Contract management guide

Source: Partnerships UK 2010.

Pre-1997 Government policies

The PFI was initially slow to start – in 1993 and 1994 only three projects, which involved over £5 m of capital expenditure, were signed. A 'Private Finance Panel' was set up in 1993, and the UK Government took a view that that PFI should be considered for any public sector project (the 'universal testing rule').

In 1996, a Government inquiry considered a number of issues, including whether PFI spending was extra or in substitution for government spending; whether the private sector would be setting priorities between schemes; whether the implications for future public expenditure were being suitably controlled; whether, and if so how, better value for money would be achieved; whether it was sensible to consider all projects for PFI; and on the specification of outputs and transfer of risks. As a result of this, the Government undertook that future spending implications of PFI would be listed in the Financial Statement and Budget Report, said that value for money gains were expected from close integration of services with design, better allocation of risks and the correct incentive structure; and identified cases where PFI would not be appropriate.

At the beginning of the 1997 Parliament the new Government abandoned the 'universal testing rule' and commissioned Sir Malcolm Bates to review the system of PFI. As a result of accepting the recommendations of the review the Government abolished the Private Finance Panel and replaced it with a Treasury Taskforce, consisting of two 'arms':

1. A *policy arm*, responsible for rules, procedure and best practice governing PFI and PPPs, together with PFI-oriented staff training of public sector employees

2. A *projects arm*, to approve ('sign off') the commercial viability of all significant projects before the procurement process began (by publishing a contract notice in the EU Official Journal) and monitor them (and other projects, where time and resources permitted) to ensure progress. The Treasury later defined 'significant project' as 'big, high profile, highly replicable or ground breaking' and the Taskforce undertook to monitor 80 such projects. Local authority projects are signed off and monitored by the *Project Review Group*, which is chaired by the Taskforce and also contains representatives of Public Private Projects Panel Ltd (the '4Ps'), an adviser to local government established by the local government associations in April 1996.

The Bates Review recommended that individual departments should remain responsible for their own PFI projects, and that departmental Private Finance Units (PFUs) should be strengthened by the appropriate expertise. As a result, the Taskforce Projects Arm was not expected to be needed indefinitely, and it was set up with a life of two years (until late 1999).

Post-1997 Government policy

As the initial term of the Taskforce came towards its end, the Government announced in November 1998 that Sir Malcolm Bates would conduct a second review: the report was made in March and published in July, 1999. Its principal conclusion was that centralised project support was still needed but that the Taskforce Projects Arm should be replaced by a joint public-private sector body, subsequently named Partnerships UK (PUK). The role of PUK has been explained in further detail later. Statutory arrangements for PUK are contained in the Government Resources and Accounts Bill, which had been enacted in 2000.

In parallel with the second Bates Review, the Government asked Sir Peter Gershon to examine civil procurement in central Government, and his report was also published in July 1999. It recommended that an Office of Government Commerce (OGC) should be created within the Treasury. The Taskforce would continue within the OGC, but with a 'slimmed down projects capability'.

The National Audit Office (NAO) had also been examining the early PFI projects, producing a number of reports, which had given rise to corresponding reports by the Committee of Public Accounts (PAC), which also made a general report drawing together its previous recommendations. The NAO had also made a report setting out the factors that it will take into account in future assessments of PFI projects. Particular emphasis was placed on the value for money (V*f*M) obligation in all PPP projects.

In July 1999, the Treasury Taskforce appointed Arthur Andersen as consultants to examine value for money aspects of those PFI projects where the delivery of services and payments for them had begun. In addition, the Institute for Public Policy Research (IPPR) launched a Commission on Public-Private Partnerships in September 1999, with the report being published in 2001.

In March 2000, the Government restated its policy on PPPs and PFI in a document entitled *Public Private Partnerships: The Government's Approach.* This policy document stipulated the roles and responsibilities that public sector and private sector were to abide by in terms of all PPP arrangements. What follows is a brief summary of the roles and responsibilities as set out in the March 2000 document.

The fundamental role for Government

First and foremost, while the best way to deliver the Government's objectives may be through some combination of public and private sectors, Government retains the responsibility and democratic accountability for:

- deciding between competing objectives

- defining the chosen objectives, and then seeing that they are delivered to the standards required

- ensuring that wider public interests are safeguarded

In the case of PPPs introduced into public services, this means that, while responsibility for many elements of service delivery may transfer to the private sector, the public sector remains responsible for:

- deciding, as the collective purchaser of public services, on the level of services that are required, and the public sector resources which are available to pay for them

- setting and monitoring safety, quality and performance standards for those services

- enforcing those standards, taking action if they are not delivered

Similarly, in the case of state-owned businesses, while PPPs bring the private sector into the ownership and management of the business, the Government remains responsible for safeguarding public interest issues. This includes, in particular, putting in place independent regulatory bodies, remaining in the public sector, whose role is to ensure that high safety standards are maintained, and that any monopoly power is not abused.

The contribution of the private sector

PPPs recognise that this potential is only partially released in the absence of the private sector. The private sector can expand opportunity through the following disciplines and skills.

Commercial incentives

Private sector organisations operate in a fluid and fast moving environment. If they do not generate profitable business, they will not survive. The realities of the private sector market-place exert a powerful discipline on private sector management and employees to maximise efficiency and take full advantage of business opportunities as they arise.

These disciplines can never be fully replicated in the public sector since there are a multiplicity of policy objectives, and a more risk-averse culture driven in part by the desire to safeguard taxpayers' money. Compared with the private sector, therefore, the public sector can be less equipped to challenge inefficiency and outdated working practices, and to develop imaginative approaches to delivering public services and managing state-owned assets.

With public private partnerships, the Government seeks to harness the innovation and disciplines of the private sector, by introducing private sector investors who put their own capital at risk. This is achieved either by introducing private sector ownership into a state-owned asset or business, or by contractual arrangements in which the private sector bears the financial risk involved in delivering a particular service or other form of specified output. If the business or service provider operates in a competitive market, market disciplines will provide an incentive to maximise quality of service. If such disciplines do not exist, and cannot easily be introduced, quality standards can be enforced through regulation or by performance requirements in the contract with the public sector. By harnessing private sector disciplines in this way, PPPs can help to improve value for money, so enabling the Government to provide more public services and to a higher standard within the resources available.

A focus on customer requirements

The need for private sector businesses to generate a return means that they are forced to look for ways to enhance the service they offer their customers, and to adapt to their changing requirements and expectations. Unless they do, customers will go elsewhere. Such incentives tend to be less clear for public sector providers, so they can tend to respond more slowly to customer demands.

New and innovative approaches

Similarly, the search for new opportunities to develop profitable business provides the private sector with an incentive to innovate and try out new ideas – this in turn can lead to better value services, delivered more flexibly and to a higher standard.

Business and management expertise

The private sector is normally far more skilled in running business activities and some elements of service delivery, including managing complex investment projects to time and budget, and assessing the commercial opportunities of potential new business ventures.

The summary of this document concluded that:

- PPPs enable the Government to tap into the disciplines, incentives, skills and expertise which private sector firms have developed in the course of their normal everyday business

- PPPs enable to release the full potential of the people, knowledge and assets in the public sector, etc.

- PPPs thus enable the Government to deliver its objectives better and to focus on those activities, fundamental to the role of Government, which are best performed by the public sector – procuring services, enforcing standards and protecting the public interest

In accordance with the policies set out in their policy document, in 2000, the UK Government set up a new body, Partnerships UK, to oversee the PPP projects in UK.

Partnerships UK

Partnerships UK (PUK) was a public private partnership which would work with both the public and private sectors to address the key weaknesses in the PFI/PPP process. By working in partnership with the public sector, it would seek to make the public sector a more effective client and ensure the best possible deal for the public sector in privately financed investment programmes. In effect, it was set up to enhance the public sector's 'intelligent client' capability.

The aim of PUK is to deliver better value for money by working on the side of the public sector. For a particular project, it would align itself with the public sector procuring authority and inject more detailed examination of practical considerations into the decision making process and drive forward the conclusion of deals. In this way, and by making available its experienced development staff and resources to assist with the development of projects, it would help departments and other public sector organisations make a better job of procuring and delivering PPP/PFI deals.

PUK have no form of monopoly or guaranteed market but seeks to win business on the strength of its offer. The Government was confident that it would be good for the public sector and the private sector alike:

> for the public sector, because its activities would boost the flow of investment into the nation's infrastructure and help the public sector achieve stronger value for money purchasing in PPP/PFI deals

> for the private sector, because it would contribute to the creation of a better flow of well-structured projects and bring about a long-awaited reduction in the cost, delay and uncertainty experienced by bidders for PFI projects

By May 2010, over 920 projects have been overseen by Partnership UK.

Infrastructure UK

In 2010, the UK Government set up a second unit, Infrastructure UK (IUK), whose role is to provide a new strategic focus across a range of sectors. IUK comprises HM Treasury's PPP policy team and Infrastructure Finance Unit and the capabilities within Partnerships UK that support the delivery of major projects and programmes. IUK advises government on the long-term infrastructure needs of the UK and provides commercial expertise to support major projects and programmes. It looks across all key infrastructure networks and both the public and the private sectors to identify and address key cross-cutting issues. It is also responsible for identifying and attracting new sources of private sector investment in infrastructure; supporting HM Treasury in prioritising the Government's investment in infrastructure; and helping to build stronger infrastructure delivery capability across government.

The European Commission guidance on PPP

In general terms, there is no special legal or statutory framework for PPP/PFI procurement. However, where applicable, EC public procurement rules and in particular the EC Public Sector Procurement Directive 2004/18 (as incorporated into English law), in addition to the general EC Treaty principles have to be followed where relevant.

3.15 Briefing Note

European Commission PPP green paper: 2004

In April 2004, the European Commission issued a green paper entitled 'On Public Private Partnerships and Community Law on Public Contracts and Concessions'.

The term public–private partnership ('PPP') is not defined at Community level. In general, the term refers to forms of cooperation between public authorities and the world of business which aim to ensure the funding, construction, renovation, management or maintenance of an infrastructure or the provision of a service.

This Green Paper analyses PPPs with regard to Community law on public procurement and concessions. Under Community law, there is no specific system governing PPPs. PPPs that qualify as 'public contracts' under the Directives coordinating procedures for the award of public contracts must comply with the detailed provisions of those Directives. PPPs qualifying as 'works concessions' are covered only by a few scattered provisions of secondary legislation and PPPs qualifying as 'service concessions' are not covered by the 'public contracts' Directives at all. Nevertheless, all contracts in which a public body awards work involving an economic activity to a third party, whether covered by secondary legislation or not, must be examined in the light of the rules and principles of the EC Treaty including in particular the principles of transparency, equal treatment, proportionality and mutual recognition.

The aim of the Green Paper was to explore how procurement law applies to the different forms of PPP developing in the Member States, in order to assess whether there is a need to clarify, complement or improve the current legal framework at the European level.

It describes the ways in which the rules and the principles deriving from Community law on public contracts and concessions are applied when a private partner is being selected, and for the subsequent duration of the contract, in the context of different types of PPP. The Green Paper also asks a set of questions intended to find out more about how these rules and principles work in practice, so that the Commission can determine whether they are sufficiently clear and suitable for the requirements and characteristics of PPPs.

European Commission PPP communication: 2005

Following the public debate on the PPP Green Paper, in November 2005, the Commission adopted a Communication on PPPs and Community Law on Public Procurement and Concessions. This Communication presents policy options with a view to ensuring effective competition for PPPs without unduly limiting the flexibility needed to design innovative and often complex projects.

Guidance on institutionalised PPPs: 2008

In February 2008, the Commission adopted an Interpretative Communication on the application of Community law on Public Procurement and Concessions to Institutionalised Public-Private Partnerships (IPPP).

The Communication explains the EC rules to comply with when private partners are chosen for IPPP. Depending on the nature of the task (public contract or concession) to be attributed to the IPPP, either the Public Procurement Directives or the general EC Treaty principles apply to the selection procedure of the private partner. The Communication expresses the view of the Commission that under Community law one tendering procedure suffices when IPPP are set up. Accordingly, Community law does not require a double tendering – one for selecting the private partner to the IPPP and another one for awarding public contracts or concessions to the public-private entity – when IPPP are established.

The Communication also states that as a matter of principle IPPP must remain within the scope of their initial object and cannot obtain any further public contracts or concessions without a procedure respecting Community law on public contracts and concessions. However, it is acknowledged that IPPP are usually set up to provide services over a fairly long period and must, thus, be able to adjust to certain changes in the economic, legal or technical environment. The Communication explains the conditions under which these developments could be taken into account.

Spread of PPP projects

The value, range and scope of the PPP projects have grown in multitudes in last decade or so, with the Channel Tunnel Rail Link being the highest value PFI project awarded in UK so far. While there still remains a significant portion of the public sector procurement on non-PPP routes, PPPs have been a significant route of public sector procurement. Table 4.1 summarises the commitment value on PPP/PFI projects (although the phrase had not been coined at that stage) during the period between 1986 and 1998.

Since 1998, the growth and spread of PPP/PFIs have been to the extent that other countries have been looking to UK when seeking ways of getting the private sector to help in the development of public sector services. A host of countries including Holland, South Africa, Portugal and Finland have expressed interest and have been looking at the UK model.

The Office of Government Commerce (OGC) reported in December 2002 that a total of more than 500 PFI projects have been signed to date in the UK. In excess of 80% of new hospitals were being built using private finance.

PPP/PFI projects: Current economic climate

The economic environment since 2008 have created some additional caveats and stipulations with respect to PPP/PFI projects.

In the past, the public sector would have expected:

- bidders to be in discussion with potential funders even if formal due diligence is conducted after the appointment of preferred bidder

- terms from funders to be reflected in bid models with a credible statement from funders that terms are likely to be maintained until financial close

- bidders to offer significant certainty about the availability and terms (including pricing) of debt

- bidders to manage the relationship with funders (including satisfaction with key contractual positions adopted such as subcontracts) so that this is not a significant interface for the public sector

However, in the current market, bidders have found that it is difficult to offer certainty about the availability and terms of funding, or – where they have felt able to offer such certainty – problems have nonetheless arisen. The overall effect is that funding remains less certain until later in the procurement and the public sector's risk of adverse affordability and/or contractual changes increases.

It is the responsibility of the authority (i.e. the public sector client) and its advisers to determine how much funder involvement is required prior to selecting the preferred bidder. Authorities should work with their financial advisers to realistically assess their projects so as to strike a balance between unnecessarily making demands on funders where the risk profile does not merit it and failing to identify aspects that may cause delay later due to funder concerns. Project factors to consider include complexity and, potentially, size

- For smaller, straightforward projects (e.g. those without demand risk, significant retained estate, challenging construction requirements, unproven technology or high operational gearing or non-standard capital contributions), funder engagement should generally remain limited at this stage. Instead, authorities and their advisers should therefore focus on ensuring bids are financeable prior to selecting the preferred bidder.

- For complex projects, authorities should satisfy themselves that funders are aware of the complexities and have not raised any issues that affect financeability.

- For large projects where the banking club involves a significant proportion of the available funders, more involvement is likely to be appropriate and authorities should satisfy themselves that the terms assumed represent an appropriately conformed position for this stage of dialogue.

Once authorities have determined the appropriate level of funder engagement, they should apply the chosen approach consistently for all bidders.

Where authorities do seek funder involvement prior to the selection of the preferred bidder, funders should be asked to identify any aspects of the project that they believe would materially adversely affect their ability to obtain credit approval. Authorities should understand exactly what areas of the documents, financial modelling, contractual risk allocation (including subcontracts), and technical aspects of the project the funders have considered in reaching their conclusion. Vague support letters stating that the project documents, preliminary technical report and financial model have been made available to funders should be discouraged in favour of disclosure of specific work undertaken by funders. As such, funders should explicitly acknowledge in their support letters the subcontract flow down inherent in the bid price and disclose any side letters or other conditions that have been communicated to sponsors. Sponsors should have a parallel obligation to disclose such side letters or conditions as part of their bid. Where funders have supported the bid, these representations should be repeated in the preferred bidder appointment letter by the selected bidder and the funders.

The objective at this stage, from the public sector perspective, is not to have sponsors eliminate funders based upon their responses, but rather to allow the authority to have early warning about difficult issues that would have arisen in any case later in dialogue or post preferred bidder.

The prevailing Government advice is that the authorities should reject a funding proposal that relies upon the use of public sector finance at this stage of procurement. Bidders should not submit bids including public sector funding and authorities should not assume that public sector funding will be available at the projected financial close date.

In an environment in which bank lending teams are resource constrained, it is preferable that funders are not asked to carry out significant early due diligence on projects that do not present unusual technical, legal or financial risks. However, for more complex projects, authorities should be aware of issues which might affect funder appetite for the project or give rise to higher financing costs, either directly through the credit margin or indirectly through performance bonding, reserve requirements and coverage ratios in excess of those required on plain vanilla projects. Authorities should seek advice from their financial advisers in making this determination. To avoid later difficulties with funders, it is essential that this assessment be made conservatively. Factors that might concern funders include:

- site with known adverse ground conditions or restricted accessibility

- high operational gearing

- substantial corporate support behind key subcontracts

- significant refurbishment works with a large number of 'unknowns'

- requirement for novel or unusual construction techniques

- construction period of more than three years

- construction materials or equipment with a unique supplier

- fixed or unusually rigid construction programme delivery dates

- assets that will be difficult to insure

- demand or volume risk

- unproven technology

- training/educational outcome risk

- unusual maintenance requirements

- services requiring a unique supplier or very limited number of suppliers

- funding solution that requires novel intercreditor arrangements

- timing & conditions of authority capital contributions

- project with multiple authorities

Authorities should be confident that the contractor support package and overall risk allocation will be acceptable to funders.

The reduction in the number of banks participating in the PPP financing market has increased the number of projects for which the remaining lenders are considering providing funds. Faced with market uncertainty, these institutions may have not significantly increased their staffing numbers with the result that lending teams may be committed to more projects than they can manage within the timetables for those projects.

Risk

The relationship between risk, value for money and affordability is one of the most complex and controversial areas of the PPP/PFI. In simple terms, the Outline Business Case (OBC) should identify all the risks associated with the project and how they are to be allocated between the purchaser and the providing consortium. Bids are then invited from the private sector to operate the scheme according to the suggested risk allocations. A Public Sector Comparator (PSC) is calculated, showing the cost of the scheme if it were to be created and managed wholly within the public sector, and based on the assumption that all the risks associated with the scheme are borne by the purchaser.

If the PSC is more expensive than the private sector bid, it is an indication that the PPP/PFI scheme will offer value for money, but the purchaser must still demonstrate that the scheme is affordable – that it has the resources to commission and pay for the scheme's long-term operation. Because the client is typically committed by contract to make payments over 30 years, affordability is a crucial issue.

3.15 Briefing Note

The allocation of risk

All organisations face the risk that outcomes will differ from those planned. Risks arise as a result of uncertainty about the future, but also from inaccurate information about existing conditions, or as a result of the failure of systems designed to contain or control risk. All individuals and organisations deal with risks every day and while some, such as the risk of a train strike preventing a planned meeting, may have relatively minor implications, others can give rise to significant additional costs.

All projects in all sectors involve risks, but the introduction of the PPP/PFI has caused public sector bodies to focus on risk to an extent unknown under traditional methods of procurement. Risks should be borne by the party best able to control or manage them, but purchasers must also consider the need for appropriate risk transfer to the private sector.

There is generally little scope for purchasers to transfer to the private sector risks arising from changes in demand for the services provided from PPP/PFI assets. Risks linked to the design, construction and operation of PPP/PFI assets can more easily be allocated to the private sector, or shared between the purchaser and the private sector – often, but not necessarily, at an extra cost – these have been further discussed later in this section.

Experiences of PPP projects tend to confirm that risk allocation did not alter much between the production of the OBC and contract close. Purchasing authorities stated their own requirements in terms of risk and outline affordability, and private sector contractors largely delivered proposals that matched, rather than sought to change, these requirements.

The reduction of risk

The emphasis on risk in the PPP/PFI gives purchasers the opportunity to think creatively about how the cost of risk can be reduced. For example, a purchaser may impose a requirement on a provider that a lift is guaranteed to be operating for the entire working day, every day of the week. This creates a high operating risk for contractors, the cost of which will be passed on through the unitary charge. Purchasers could explore other possibilities with the contractor to reduce the potential cost of that operating risk. The contractor may be able to make other space available, temporarily, in the case of a lift breakdown, effectively reducing the significance of the lift's reliability in the risk model and on the cost of the project. We found little evidence, however, that purchasers were taking these more creative approaches to risk reduction.

Government guidance states that purchasers should not transfer risks to the operator to get a particular accounting result if this arrangement delivers poorer value for money. In practice, however, there is not always a linear relationship between the levels of risks accepted by the contractor and the price attached to these in a contract. The strength of the link is likely to be influenced by the state of the market and the profitability of the whole contract.

Moreover, large contractors, using their own equity, do not have to convince financiers that the risks involved are reasonable and justifiable. If contractors really want to get involved, they are likely to be prepared to accept a package of risks, and clients may be surprised at the apparently low price attached to some risks by contractors, compared with their own estimates. The result is that risk transfers from the purchaser to the contractor do not necessarily mean significant extra costs and reduced value for money.

Quantified risks, and associated probabilities, could be used more often to find out whether certain risks allocated to the contractor should be re-assumed during negotiation. If the contractor were to be asked how much the price would be reduced if a risk was reallocated to the public sector, value for money could be improved.

Incomplete use is currently made of risk models in exploring risks and their probabilities. Instead, risk models still tend to be used as 'necessary' number-crunching exercises to demonstrate whether the PPP/PFI scheme is better than the conventionally funded alternative.

While risk models can be used to assist skilled negotiators by showing general tendencies in the likely movement of price with risk, they cannot be relied on to generate a 'correct' answer. Given the nature of the market, and the approach adopted by individual companies, PPP/PFI negotiation around risk is rarely simple or predictable.

The transfer of risk

With the PFI, as with many other types of PPPs, value for money is achieved through the transfer of risk to the private sector, which is perceived to have an advantage in handling risk. The risks that can be transferred to the private sector can be divided into two groups, *general risks* that are common all types of public/private service projects and PFI *specific risks* that are PFI public services project specific.

PFI specific risks

Risks come in many forms and often depend on the characteristics of a particular project. The risks involved in providing a playground for a school are sure to differ in some aspects from the risks associated with a large-scale transport project. The transfer of risk differs with PFI public services contracts in that it enables the transfer of project financing risk to the private sector. Therefore, an essential condition of any PFI project is that sufficient financial risk is transferred to the private sector to secure value for money. The main benefit of transferring financial risk to the private sector is that they are perceived to have an advantage over the public sector in handling financial risks. Most successful private sector firms have risk analysts especially in the financial sector. Public services project financing risk; the risk of delivering an economically viable financial package, can be divided into two main types, internal *disposal risk* and external *financing risks*. *Disposal risk* is the risk that the expected value of surplus departmental assets, detailed for disposal in a PFI contract to fund public services, is lower than expected. Departments can reduce their exposure to this risk by transferring assets, such as redundant hospital buildings and grounds, which have, or are to become, surplus to requirement to the private sector contractor as part of the PFI contract. *External financing risk* is the risk that the private sector contractor fails to raise sufficient funding for a public services project on the market. As with any contract, the ability of the private sector contractor to secure the finance required to complete a PFI project, must be determined by the sponsoring department before the deal is signed. External financing risks are also related to *interest rate risk*, which is the risk that the interest rate will change between the time a bid is tendered and the time a contract is signed. Adverse movements in the interest rate during this time mean that the private sector contractor has to pay more to service their debt, which may reduce the attractiveness of a PFI contract. The transfer of project financing risk generates incentives for the private sector to supply services on time and of a higher quality as they only start to receive service payments when a flow of public services actually starts, and continued payment depends on meeting specified performance criteria.55 A further effect of transferring a project's financing risk to the private sector is that it reduces the general risks of public service projects that have been retained by the public sector. However, risk and reward go hand in hand: the higher the perceived risk that is being transferred to the private sector, the greater the risk premium that will be required by the contractor from the public sector to compensate them for their exposure. Given that some risks are difficult to quantify it is difficult to determine whether a private sector contractor, for accepting a particular risk, is charging a suitable risk premium for either party.

The optimal allocation of risk

Once the risks associated with a particular PFI project have been identified, the next task is to share the risks between the public and private partners. In keeping with the well-accepted principle that 'risk should be allocated to whoever is best able to manage it', the public sector must not risk transfer for its own sake. However, as a general rule PFI schemes tend to always transfer to the supplier design, construction and operating risks (both cost and performance). Demand and other risks should be a matter of negotiation with the value for money impact being tested out, where appropriate, through bids on alternative risk transfer bases against minimum and conforming requirements.

Risks retained by the public sector include:

- The risk of incorrect requirement specification: Where it is known that requirements cannot be specified in their entirety initially, as in some IS/IT projects, it may be possible to share with the supplier the risk of defining remaining requirements during developments and implementation. The public sector still retains the risk in respect of the initial specification.

- The risk of criticism: A failure of a public service, even if entirely the responsibility of a supplier, may result in criticism of the Government or local authority along with the supplier.

- The risk of the long term need for the service: As these contracts often run for 25 years or may be even longer, the risk that a particular service may no longer be needed remain with the public sector. Even if the service may no longer be required, the contractor will be entitled to its payments throughout the entire duration of the contract

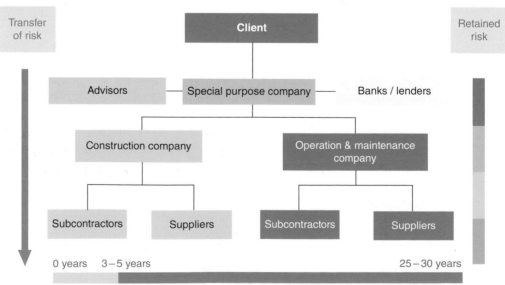

Generic risk transfer model in PPP/PFI projects.

The risk of long term authority affordability of the project

The risk of land acquisition

Discriminatory or specific changes in Law

The risks of a public services project should only be transferred to the private sector if, and to the extent that, the private sector is capable of managing such risk. In situations where the private sector is best judged able to deal with risk, such as construction risk, then the public sector should try and transfer this responsibility completely. Where the private sector is deemed less able to manage project risk, responsibility for these risks should remain within the public sector.

The following figure highlights a generic risk transfer model for PPP/PFI projects.

Planning

Risks associated with planning approval and related issues generally lie with the public sector. In some instances, particularly in the transport sector, the Government may use statutory authorisation which would negate the necessity for using any general planning law, thus avoiding this risk.

The UK planning process is not tailored to PPP/PFI projects and can often be quite slow, with the outcome being uncertain. It is therefore important for the public sector to factor this in within the risk matrix.

When the procuring authority is a local authority, there needs to be sufficient and reasonable separation between the functions of the local authority and the planning authority for pre-empt any future judicial review challenge on the grounds of conflict of interest.

Risks associated with Environmental Impact Assessments, Public Consultations and Stakeholder Consultations usually rest with the public sector.

While it is usually a public sector risk to provide the land for the project, in certain circumstances it is possible to transfer this risk to the private sector, for example, making it a tender requirement; If a site needs to be acquired for a PPP/PFI project and negotiation with the land owner proves to be unsuccessful, in certain circumstances public authorities may be able to obtain the power to acquire the land compulsorily. However, the process for this is not tailored for PPP/PFI projects and can be quite slow, with the results being uncertain.

Design

Quite simply almost all the design risk is transferred to the private sector in a PPP/PFI arrangement. However, the risks for inadequate or incorrect initial specification or change in authority requirements after tender/award remain with the public sector. In some projects, for example in IS/IT projects, it may not be possible for the public authority to specify exact requirements at the tender stage or award stage. In these circumstances, risk share is usually negotiated between the parties.

Construction

The constructions risks, broadly speaking, lie with the private sector in almost all the PPP/PFI arrangements. This is in line with the widely accepted wisdom that risks should be transferred to those who are best placed to manage them.

However, should the initial specification be incorrect or inadequate, or should there be specification variations, any construction risks arisen thereof will have to be analysed and allocated in accordance with the terms of engagement.

3.15 Briefing Note

As is the case in the construction industry, the construction related risks are usually transferred down the chain to the subcontractors. However, where liability is being passed by the project company, there may need to be a further mechanism to apportion it between subcontractors. This can be addressed in an interface agreement between the subcontractors.

In a PFI project, these are commonly 'equivalent project relief' provisions in the subcontracts, which seek to match the claims of the subcontractors against the project company with equivalent claims of the project company against the public sector authority. However, in a recent case, the Court held that under English legislation, a construction subcontractor could not be prevented from referring disputes immediately to adjudication and that certain of the particular 'equivalent project relief' provisions in the construction subcontract in question were ineffective. Although the potential for mismatch between decisions at the subcontract and project agreement levels is attempted to be mitigated by the use of these 'equivalent project relief' provisions, legislations contribute to the danger of inconsistent decisions as construction and maintenance subcontracts are required to allow disputes to be referred to adjudication while PPP/PFI project agreements are exempted from this requirement.

PFI/PPP creditors are rarely willing to accept the raw construction risk that the project does not get built on time and on budget. So, the Project Company will – in addition to structuring the construction agreement on generally more restrictive terms than the obligations in the project agreement – either require the Contractor provide a construction risk mitigation package or arrange one itself and pass some of the costs on to the Contractor. In some jurisdictions, regulations also specify minimum levels of external support that must be provided in order to be eligible to bid on government contracts.

The design of a construction risk mitigation package always begins with consideration of the stand-alone risk of construction in a given project. It then incorporates any of several possible financial and performance supports. The supports are designed to shield the project company and ultimately lenders from cost or schedule over-runs on the project and are also intended to raise the credit quality of the project debt during construction.

Operating risks

The allocation of operating risks will very much depend on the nature, scope and context of the PPP project. For example, the operating risk allocation of DBFO contract will be different to that of an operate-only contract.

In PPP/PFI agreements, staff are often transferred to the employment of the private sector, in accordance with the Acquired Rights Directive 77/187 enacted by the Parliament as Transfer of Undertakings (Protection of Employment) Regulations (more commonly known as TUPE). There are a number of risks associated with employee (and pension) transfer, including the risk of suitability, conduct, performance and control of staff, risk of short fall in the pension, risk of failure to comply with TUPE and unforeseen liabilities arising out of TUPE, etc.

In addition, risks such as under-performance due to incorrect specifications, technological obsolescence and change in performance criteria remain with the public sector.

Demand risks

This is the risk that demands for the asset will be greater or less than predicted/expected. Where demand risk is significant, it will normally give the clearest evidence of who should record an asset on their balance sheet. For example, the demand for hospital beds by patients may be less or more than what was predicted.

The length of the contract may influence the significance of demand risk since it is difficult to forecast for later periods. Once it is established that demand risk is significant, it is necessary to determine who will bear it.

The importance of demand risk is linked with the financial arrangements that are tied to the demand prediction. Hence, risks associated with any changes to the scope of demand will usually rest with the public sector.

If the revenue generation is directly linked with the operation and part of the contractor payment mechanism, any shortfall in revenue generation due to change in demand will normally be an operator risk. On the other hand, the risk of fall in demand due to a change in the Government policy, political decision, social, economic or environmental change usually will remain with the public sector.

Financial risks

As previously discussed, it is in the interest of the public sector to transfer the financial risk to the private sector in PPP/PFI arrangements; however, there are some exceptions where the financial risk lies with the public sector. These include insufficiency of the public funds or ability to pay over the contract duration (e.g. for a period of 25 years). Furthermore, if there is a requirement for the public sector for off-balance sheet treatment, that also has an impact in risk allocation.

Usually the currency if finance tends to be pound sterling, and if the source of finance is international, the finance tends to be sourced from a UK branch of the relevant financial institution. In any case, the foreign exchange fluctuation risks, if applicable, lies with the private sector.

Risks for other finance criteria, where private finance is utilised, usually stay with the private sector, including change in taxation (unless it is a discriminatory or specific legislation), insurance and finance arrangements such as equity and bond.

There are certain circumstances, for example, when the PPP/PFI arrangement may be exempt from Stamp Duty Land Tax; however, the public sector, should it decide to take advantage of this, has to ensure that the appropriate structures and conditions are met as PPP/PFI projects inherently may not conform to the exemption requirements.

Legislative risks

The allocation of legislative risks would normally depend upon whether it is a general change in law as opposed to a discriminatory or specific change in law.

Changes in law which are generally applicable are normally a risk for the private sector, with some notable exceptions. If there is a general change in law which comes into effect during the service period and involves capital expenditure, there is often a sharing of risk, with the exposure for the private sector contractor to such capital expenditure being on a sliding scale, with a capped value for the contractors total exposure. Similarly, risk of changes in VAT status of the contractor is also an exception which is normally protected against.

On the other hand, the public sector generally retains the risk of any changes in law which would expressly discriminate against the PPP/PFI project, the project contractor or the PFI sector. Similarly, changes in law which specifically refer to the construction and servicing of facilities for the sector in question is a public sector risk, providing such a change would not have been foreseeable at the time of the project agreement.

3.15 Briefing Note

Residual risks

Residual risk is the risk that the actual residual value of the asset at the end of the contract will be different from that expected. The risk is more significant the shorter the PPP/PFI contract is in relation to the useful economic life of the asset.

Where this risk is significant, who bears it will depend on the arrangements at the end of the contract. For example, the public sector will bear the residual value risk where:

- it will purchase the asset for a substantially fixed or nominal amount at the end of the contract

- the property will be transferred to a new private sector partner, selected by the public sector, for a substantially fixed or nominal amount

- payments over the term of the PFI contract are sufficiently large for the private sector not to rely on an uncertain residual value for its return

On the other hand, the private sector will bear residual value risk where:

- it will retain the asset at the end of the contract

- the asset will be transferred to the public sector or another private sector partner at the prevailing market price

Briefing Note 3.16   Guidance on e-procurement

e-Procurement and Europe

Some of the key e-procurement participant countries in Europe include Austria, Belgium, Denmark, France, Germany, Ireland, Italy, Norway, Portugal and Sweden. There are many successful e-procurement solutions within Europe. However, at time of going to print, there is no single government organisation in Europe that has implemented a comprehensive suite of electronic tools and systems to support all public procurement activity. Comparatively, the UK government has made admirable progress in this area. Considerable investment and commitment to e-procurement continues across Europe.

EU Directives

The EU encourages the use of e-procurement. The new EU Consolidated Directives and EU Invoicing Directives make clear provision for the use of electronic tools and techniques within public sector purchasing across Europe. The EU acknowledges that automating processes and enabling opportunities to be advertised and tendered online fully supports the aims and objectives of cross-border trading, non-discrimination and fair and open competition. They are also encouraged by the transparency and the ease of auditing online transactions. Innovative tools such as electronic reverse auctions are provided for within the new Directives, and the EU has even gone a step further by introducing a new process, Dynamic Purchasing Systems, an online-only process whereby suppliers can compete for contracts.

e-Procurement best practice

The 'quick wins' approach

To modernise procurement processes, certain 'e-tools' will be required. One problem often encountered at the start of an e-procurement programme is deciding which tools to implement and in what order. 'Quick wins' can establish the credibility of the e-procurement programme, and help to generate funding for the rest of the programme. Examples of these can include use of procurement cards [similar to the Government Procurement Card (GPC)], e-auctions, e-sourcing and similar initiatives.

Implementing e-procurement

A wide variety of e-procurement tools have been developed over recent years to help organisations source, contract and purchase more efficiently and effectively. Broadly, e-procurement tools relate to two aspects of procurement: sourcing activity and transactional purchasing.

Sourcing activity (e-sourcing)

The e-sourcing tools can help buyers establish optimum contracts with suppliers and manage them effectively. The tools include supplier databases and electronic tendering tools, evaluation, collaboration and negotiation tools. Also included are e-auction tools and those tools which support contract management activity.

Transactional purchasing (e-purchasing)

The e-purchasing tools can help procurement professionals and end users achieve more efficient processes and more accurate order details. The two aims of maximising control and process efficiency are the function of e-purchasing tools such as purchase-to-pay systems, purchasing cards and electronic invoicing solutions. Although the tools fall broadly within these two categories, some tools can be implemented in isolation.

e-Auction tools are now a mature technology that can generally be implemented more quickly than other e-sourcing tools. As e-auctions are currently proving a clear 'quick win' in cash-releasing terms, their earliest implementation is strongly recommended. The following diagram shows where e-sourcing and e-purchasing (called purchase to pay) tools fit in the procurement lifecycle as classified by the Chartered Institute of Purchasing and Supply (CIPS).

Purchasing cards (P-cards)

Purchasing cards (P-cards) are similar in principle to charge cards used by consumers (e.g. suppliers are paid within five days; the buyer is billed monthly in a consolidated invoice), but with extra features which make them more suitable for business-to-business purchasing. These can include: controls such as restricting card use to particular commodity areas, individual transaction values and monthly expenditure limits. The purchasing information provided to the buying organisation by an issuing bank on each monthly statement depends on the degree of detail automatically generated by each supplier. This can range from the supplier name, date and transaction value, to line item detail against each item ordered, free text entry for the input of account codes and VAT values.

Supplier participation in P-card programmes

Many suppliers already accept consumer credit and debit card payments and no extra equipment is required to accept P-cards. The costs to suppliers in accepting credit, debit and P-card payments are a small transaction charge (normally ranging from 1 to 4%) and the cost of implementing the card-processing equipment. This cost increases with the higher level capabilities.

CIPS e-procurement lifecycle.

Benefits of P-cards

- Process savings
- Prompt payment discounts
- Guaranteeing prompt payment
- Increased compliance with contracts

e-Auctions

In an electronic reverse auction (e-auction), potential suppliers compete online and in 'real time', providing prices for the goods/services under auction. Prices start at one level and gradually, throughout the course of the e-auction, reduce as suppliers offer improved terms in order to gain the contract. e-Auctions can be based on price alone or can be weighted to account for other criteria such as quality, delivery or service levels.

e-Auction benefits

Government e-auction activity shows an average saving of 13.4% over previous contract value. Further benefits include improved preparation and planning for the tendering process, opportunity for suppliers to submit revised bids for a contract (as opposed to the formal tendering process), and increased market knowledge for buyers and suppliers. Suppliers particularly benefit from increased awareness of competitor pricing.

Implementing e-auctions

e-Auctions do not replace tendering: they are a part of it and provide cost-effective, fast and transparent conclusions to a full tendering process. e-Auctions may be based on securing the lowest price, or on most economically advantageous bid (price, payment terms, supply schedules). Only those suppliers who have successfully pre-qualified (i.e. they have satisfied all tendering criteria such as quality processes, financial stability and environmental policies) should be invited to participate. The complexity of an organisation's procurement will affect the e-auction strategy. Some basic considerations for all requirements, whether complex or simple, are:

- Starting price: What will be the starting price criteria? For example, an indicative price submitted by suppliers in an earlier stage of the tendering process?
- Bid decrements: What will be the minimum level by which a supplier can reduce their bid below the current lowest? For a £100,000 contract, a bid decrement of £2000–5000 would be reasonable.
- Duration: What will be the duration of the event?
- Extensions: What extensions will be granted? For example, if any bids are received within the last five minutes of the e-auction an extension of five minutes might be granted for other bidders to respond.
- Weightings: More complex e-auctions will allow suppliers to revise their bids in respect of criteria including, but not restricted to, price.

Further information can be available from various sources including SIMAP. Also advice must be sought from the individual service providers.

Suppliers and e-auctions

Suppliers are generally cooperative about participating in e-auctions. Buyers should maintain excellent communications with suppliers, being open and providing all relevant operational and technical information (this is a legal requirement). Buyers should also provide supplier inductions to e-auctions and a test e-auction prior to the live event if necessary to ensure supplier familiarity with both the process and the technology.

Each supplier may adopt a different strategy for participation: some bidding lower prices early, others holding back. Suppliers may wish to see where they stand in the bidding process and the value of other bids, but not the names of other bidders.

3.16 Briefing Note

Use of e-auctions in the construction industry

The UK public sector is implementing e-auctions as a valuable tool for improving the purchasing process. There is now more experience in both the public and private sectors to demonstrate that e-auctions have been found to improve professionalism, speed up the process and, in many cases, reduce the purchase price for goods and services. Within the UK public sector, e-auctions are being implemented in accordance with best practice, supported by a professional code of conduct. e-Auctions form only one stage of a full-quality tender process. e-Auctions themselves can be complex with weighted options to take account of factors other than price throughout the process. This approach ensures that contracts continue to be awarded on a value for money basis and not on price alone in line with government policy.

The construction industry has been progressing with adoption of e-commerce in the same way as other sectors. However, there have been some strong objections to electronic reverse auctions (e-auctions) from some sections of the industry. Government has received representations from trade associations and other bodies. Sections of the industry have seen e-auctions as a return to lowest price purchasing, threatening already low margins. The industry also perceives e-auctions as challenging the principles of the Achieving Excellence in Construction initiative, such as an integrated supply chain approach to construction procurement based on optimum whole-life value.

3.16 Briefing Note

Briefing Note 3.17   Design management process

A framework for design management

Managing design successfully can be accomplished with a core set of tools, and this process of management has certain basic characteristics, or 'hallmarks', irrespective of the project type, size, sector, business or organisation.

Teams are brought together to design and construct something for the client. There are a number of parties involved, information needs to be produced and exchanged in various forms, suppliers, manufacturers, constructors and other specialists need to be engaged in the process of madeliver the end product, at the right time, for the right cost and providing and meeting the right quality.

How do we know that design management is taking place and is working? What is the proof? This generic framework (DMTCQ) defines activities and processes that should be in place, and can be checked.

> If design is happening, and is being managed effectively, then certain key things need to be happening as a consequence and the results of those management activities should both be visible and tangible in some shape or form.
>
> There should be visible outputs or outcomes – 'the Hallmarks of DM'.

These principles apply equally to a road, a ship, a nuclear power station, a school or a house.

This brings together several tools and processes, which should enable control of the design process. Conversely where a design management process is not delivering the results as expected, then it is possible to diagnose what and why it is not working and get the process back on track.

There is no magic or silver bullet for DM. Methodical and consistent application of strategy, management, delivery and definition of objectives and outcomes, will pay dividends. Good DM is simply good management practice and the use of appropriate techniques.

Note these are only principles and this is not prescriptive about the exact 'How' as this will vary from project to project, depending on sector, size, complexity and time scale. For example, a programme of activities and milestones could be produced as a bar chart, or a schedule of dates. What is important is that the design manager or the person taking that function, is managing the particular aspect, using an appropriate tool, and therefore able to demonstrate effective management.

3.17 Briefing Note

213

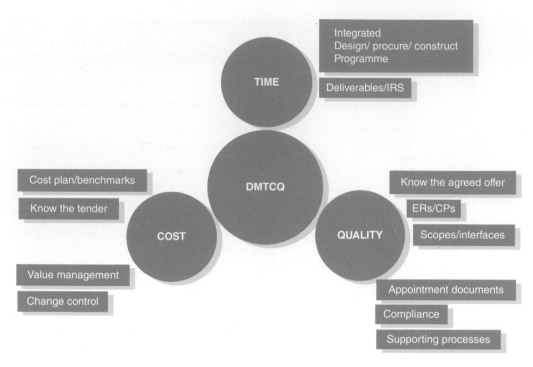

DMTCQ – a framework for design management.

DMTCQ (design management – time cost quality)

All of the tools and processes need to be in place and managed effectively to provide 'the Hallmarks of DM' that were discussed earlier.

Time

- **Program** – design/procure/construct program – Fully integrated, and agreed with the whole team. Monitored, reviewed and managed.

- **Deliverables schedules/IRS** – In place and agreed with the whole team. Monitored, reviewed and managed.

Cost

- **Cost plan/benchmarks** up to date with the design iterations, reporting in place.

- **Tender** – Know the full details of the agreed offer including any clarifications/exclusions – understand any particular conditions.

- **Value management** – Are there any potential ways of improving the value proposition of the scheme for the client? Consider design, delivery, lifecycle.

- **Change control** management procedure – TCQ implications of any design change, and to be formally reviewed and instructed.

Quality

- **Agreed offer** – Know the details of the agreed contract offer – understand what has been agreed to in terms of function, content, standard, cost and timescale.

- **ERs/CPs** – Employer's requirements/contractor's proposals – understand the relationship between these documents and the requirements.

- **Scopes/interfaces** – Understand the scope, content and interfaces of the Works Packages, check for gaps, overlaps, duplications, inconsistencies.

- **Appointment documents** – Understand exactly what the consultants are contracted to do, is everything covered? Understand their resources and capabilities. Are their services and resources sufficient and appropriate?

- **Compliance** – Planning, listed/conservation consents, Building Regulations, other code requirements, processes in place to achieve approvals.

- **Supporting processes** – In place for CDM/safety/sustainability/environment/project specific.

And remember

- **Questions** – Always have clarity of role, objectives, scope, accountabilities, for all parties. Think of the who, what, when, how, why?

- **Information** – Always know all the design information – keep the drawings, specifications, schedules, information, under constant review.

- **Risk schedule** – Be aware of the risk management processes, and ensure that risks are eliminated, reduced and managed effectively.

- **Meetings/communications/reports** – Make sure that communications and reporting are working effectively at all levels.

- **Dynamics** – Think about people issues. Are there personality issues/agendas driving situations?

- **Training** – Do any parties/individuals need training on processes, technology or other project aspects?

- **Stage/gateway/interim reviews** – Review the work at key milestones and stages, identify activities and outputs that are not complete or not up to the standard you need, have a strategy in place to deal with outstanding issues before starting the next stage.

Note that this is a very brief overview of the key issues and factors and for more information on process and tools refer to The CIOB Design Manager's Handbook.

4 Pre-construction

Stage checklist

Key processes:	Design delivery process
	Technical design and production information
	Value management
	Procurement of supply chain
	Contractual arrangements
	BIM strategy
Key objective:	What do we need to build?
	How would it look like and function?
	How would we deliver it and manage it?
Key deliverables:	Design outputs
	Contractual arrangements
Key resources:	Client team
	Project manager
	Design team
	CDM coordinator

Stage process and outcomes

This stage involves implementation of the plans developed during the strategy stage to ensure all aspects necessary for the commencement of works on site have been resolved. These involve execution of the design process, obtaining of all required statutory and legal consents and the selection and appointment of the contractor(s).

Outcomes:

- fully developed design information

- fully developed tender documentation

- design freeze

- tender process

- update Health and Safety Plan and File

- resolution of all statutory and legal consents and approvals

- selection and appointment of contractor(s)

- completion of development of site logistics

- confirmation of BIM strategy

Code of Practice for Project Management for Construction and Development, Fifth Edition. Chartered Institute of Building.
© Chartered Institute of Building 2014. Published 2014 by John Wiley & Sons, Ltd.

- confirmation of readiness to commence work on site

- client approval to proceed with the construction works

Design process

Design development is rarely completely linear. It involves experiment and revisiting solutions which turn out not to be ideal. Getting design right is therefore a process which requires insight, flexibility and collaboration. Planning and managing design delivery inherits these considerations, too.

Design involves various players, and during the process the intensity of their involvement changes. For instance, the architect may have a formative role during briefing and in the phase following it, and then in the later design stages and during construction the architect's role diminishes, while the role of specialist designer-contractors will probably start late and be very intense during parts of the construction process.

Project management processes hold the key to tackling the problems associated with design delivery, and it is right that managers should accept responsibility for making sure that the required processes are identified and implemented. For example, the brief is the client's statement of requirements, but often the client cannot be expected to understand fully what the briefing process comprises. It is up to the managers involved (project manager, design team leader) to guide the client through it and to ensure the development of the design brief fully reflects the client's requirements. The project manager has an ongoing responsibility in checking that the design, as it evolves, complies with the design brief.

The design process follows the sequence set out by the RIBA Plan of Work (see Figure 0.2) with each stage being sanctioned by a formal sign-off before commencement of the following stage. An adequately run project team cannot reasonably complain of inadequate briefing, unless the client has failed to respond fully to their questions.

Managing the design delivery

The project manager will need to convene a meeting of the design team and any other consultants/advisors to review all aspects of the project to date. A dossier of relevant information should be circulated in advance. The object of the meeting will be to formulate a design management plan. The plan should at least cover:

- who does what by when

- the size and format of drawing types

- schedules of drawings to be produced by each discipline/specialist

- relationships of interdependent CAD (computer-aided design) systems

- transfer of data by information technology

- estimates of staff hours to be spent by designers on each element or drawing

- monitoring of progress and the effect of design resources expended compared to productivity achieved

- schedules of information required/release dates

- initiating procedures for design changes to be made and their effects predicted

- incorporation within the design schedule of key dates for review of design performance to check

- strategy and technical options to meet the sustainability brief

- compliance with brief

- cost acceptance

- value engineering analysis

- health and safety issues

- completeness for tender

The project manager is ultimately responsible for ensuring that there is a system in place for monitoring and controlling the production of design information in line with the agreed schedule. The project manager should convene and attend regular design team meetings to review progress and ensure that the design team are performing in accordance with their duties. Responsibility for the coordination and integration of the works, which involves input from other designers, consultants, service providers, statutory authorities and utilities, etc., normally would lie with the design team leader, which in most cases will be the architect.

Project coordination and progress meetings

To aid control of the design process, the project manager will arrange and convene project progress meetings at relevant intervals to review progress, resources and productivity on all aspects of the project and initiate action by appropriate parties to ensure that the design management plan is adhered to or establish reasons for departure and the effects of departure. In the event of a departure, the project manager should address contingency planning and mitigatory/recovery strategy and initiate recovery. Distributing minutes of meetings to all concerned is an essential part of the follow-up action.

Design team meetings

Design team meetings are convened, chaired and minuted by the design team leader. It is not essential for the project manager to attend all team meetings as a matter of course, although he normally has the right to do so. The project manager will receive minutes of all meetings and will report to the client accordingly.

Managing design team activities

Key specialist contractors may need to be involved at an early stage and managed equally with the design team (see Figure 4.1).

The project manager has several responsibilities:

- Monitoring progress, resources and productivity against the design management plan in association with the team. This is essential in view of their interrelationship. However, effective interrelationship cannot be finalised until the full team has been appointed and had time to get to grips with the project and its complexities.

- Advising the design team leader of the requirement to agree the detail and integration of the design team activities and to submit an integrated design production schedule for coordination by the project manager.

- Incorporating into the project schedule the dates for the submission of design reports and periods for their consideration and approval.

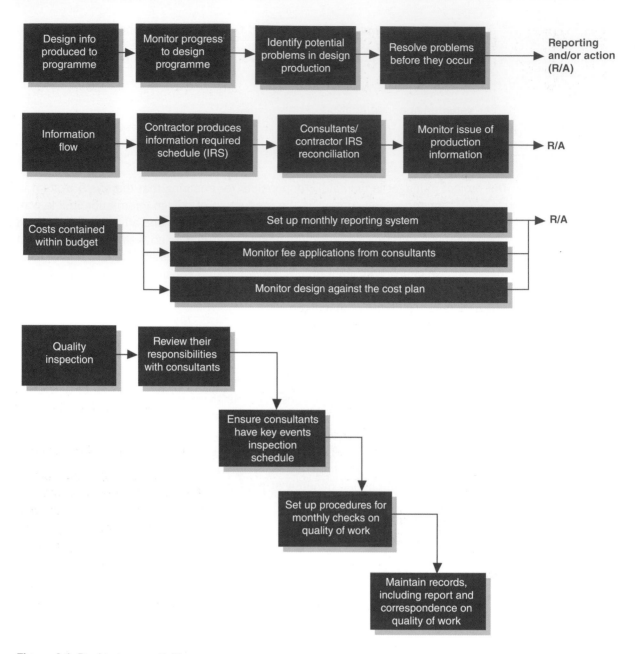

Figure 4.1 Design team activities.

- Commissioning, as necessary, or arranging for the team to commission, specialist reports, for example, relating to the site, legal opinions on easements and restrictions and similar matters.

- Ensure a competent consultant is appointed as CDM coordinator as required by CDM regulations.

- Drawing to the attention of the client and the designers their respective duties under the CDM Regulations and monitoring compliance.

- Arranging for the team to be provided with all the information it requires from the client in order to execute its duties. It is an important function of the project manager to coordinate the activities of the various (and sometimes numerous) participants in the total process. CDM coordinator, solicitors, accountants, tax advisors, development advisors, insurance brokers and others may all be involved in the pre-construction stage.

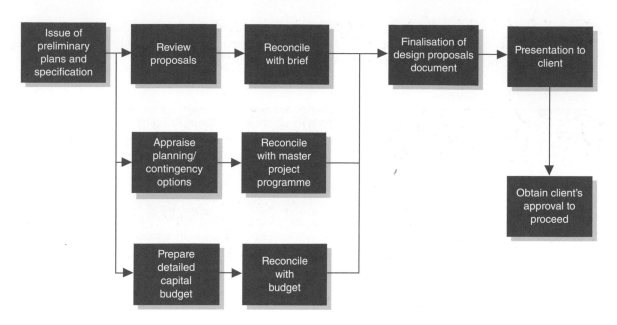

Figure 4.2 Development of design proposals.

- Submitting, in conjunction with the design team leader, design proposals, reports and design development (formerly scheme design) drawings to the client for approval (see Figure 4.2).

- Conveying approvals to the team to proceed to subsequent stages of the project.

- Obtaining regular financial/cost reports and monitoring against budget/cost plans. Initiating remedial action within the agreed brief if the cost reports show that the budget is likely to be exceeded. Solutions to problems that cannot be resolved within the agreed brief, or likely substantial budget underspend, should be submitted to the client with recommendations. The necessity to agree firm budgets at an early stage is most essential. It could, in certain cases, lead to the client modifying the project brief.

- Preparing 'schedule of consents' with action dates submission documents, status, etc., and monitor progress.

- Checking that professional indemnity insurance policies are in place and remain renewed on terms that accord with conditions of engagement.

Statutory consents

Although a great deal of the detailed work involved in obtaining statutory consents, such as planning permission and Building Regulations' approval, is carried out by the design team and other consultants, the project manager has a vital facilitating role to play in what can be critical project activities.

Planning approval

Planning consultants

On many schemes today, the complexity of the planning process is such that the project manager may appoint a planning consultant to advise on the approach most likely to secure planning consent for a scheme together with the information that will be required to support the application, including an environmental impact assessment incorporating sustainability appraisal, traffic assessment, green travel plan and ecology report. In conjunction with the other members of the design team, the planning consultant will organise and participate in key meetings with the planning officer and other departments, such as highways.

Legislation

The primary legislation governing the planning process is contained in several Acts of Parliament. The grant of planning does not remove the need to obtain any other consents that may be necessary, nor does it imply that such consents will necessarily be forthcoming.

Timing

Planning permission cannot be guaranteed or assured in advance of the local planning authority (LPA) decision within the statutory period and the project manager must recognise this by allowing a contingency in the master development schedule.

Negotiations

The project manager will normally assist the design team leader in negotiations with officers of the local authority and report to the client on the implications of any special conditions, or on the need to provide *planning gain* through the appropriate statutory agreements. The client's legal advisors are briefed to act for the client accordingly.

Presentations

The project manager will arrange, should it be necessary, any presentations to be made to LPAs and local community groups. He will also organise meetings, including agreeing publicity and press releases with the client.

Refusal

Should planning permission be refused the advice of the relevant consultants should be obtained and action initiated, either to submit amended proposals or to appeal the decision.

Appeal

In the event of an appeal, arrangements are made for the appointment and briefing of specialists and lawyers, including managing the progress of the appeal. Applicants who are refused planning permission by an LPA, or who are granted permission subject to conditions which they find unacceptable, or who do not have their applications determined within the appropriate period, may appeal to the Secretary of State. Appeals are sent to the Planning Inspectorate.

Enforcement powers

In the event of a breach of the planning legislation, the local authority's main enforcement powers are:

* to issue an enforcement notice, stating the required steps to remedy an alleged breach within a time limit (there is a right of appeal to the Secretary of State against a notice)

* to serve a stop notice which can prohibit, almost immediately any activity to which the accompanying enforcement notice relates (there is no right of appeal to the Secretary of State)

- to serve a breach condition notice if there is a failure to comply with a condition imposed on a grant of planning permissions

- to apply to the High Court or County Court for an injunction to restrain an actual or apprehended breach of planning control

- to enter privately owned land for enforcement purposes

- following the landowner's default, to enter land and carry out the remedial work required by an enforcement notice, and to charge the owner for the costs incurred in doing so

It is a criminal offence not to comply with an enforcement notice's requirements or to contravene the prohibition in a stop notice.

Other statutory consents

It is the duty of the design team to facilitate that the design complies with all other statutory controls, for example, consents for Building Regulations, means of escape, the storage of hazardous materials, fumes and emissions, and pollutants. Generally, statutory controls make the owner or occupier responsible for the aspect of continuing duties in relation to the statute. The project manager obtains from the design team and/or other relevant sources, all consents and arranges for the client to be advised of these continuing duties. Others, such as specialist subcontractors, submit and obtain Building Regulations' approval for their product/system.

Building Regulations

The Building Regulations comprise 19 approved documents (A-P), plus references to numerous EN and British Standards and other technical documents, which take up several volumes.

The Building Regulations represent the technical standards that the building must comply with, ranging from fire resistance and acoustics, to structure, means of escape and accessibility.

Ensure that advice is sought early in the design process. Achieving compliance later in the project can result in abortive design work or once on site costly changes to built work.

Building Regulations compliance will be dealt with through the Local Authority Building Control department, or by using an Approved Inspector.

Frequently a Conditional Approval is issued. This will contain a number of conditions to be discharged by the submission of further information.

In most cases this is information to be supplied by the Design Team, and sometimes by particular subcontractors.

This process of discharging conditions must be managed to ensure that conditions are discharged and the design is approved before work is ordered or installed. Otherwise the finished work will be at risk, and the Building Inspector may require changes once the work has been installed.

The simplest method of dealing this is to compose a schedule (tracker) listing:

- The conditions.

- Information required to discharge.

- Who is the owner/or lead on the condition?

- Who will produce the information?

- The schedule should be reviewed regularly at Design Team Meetings or Project Team Meetings, until completely discharged.

Disability Discrimination Act (DDA)

This Act provides for access requirements for anyone with a disability whether physical, visually impaired, aural, etc., within buildings. This goes beyond part M of the Building Regulations. Under the planning/development control process, schemes over a certain size will need an access statement as part of the information submitted for planning approval, usually completed by an access consultant.

Impact of utilities on project planning/scheduling

Due to the long lead periods required by utilities providers, that is, gas, water, electricity and communications, the project manager should ensure that the requirements in respect of diverting or increasing existing supplies or installing new supplies are identified at an early stage in the development process. The procurement of these utilities should be monitored closely to ensure that they do not affect the completion of the project. The project manager should also be aware that supplies are often required prior to construction completion to enable commissioning of the building services installations.

Technical design and production information

The project manager's monitoring and coordinating role will entail extensive liaison with members of the project team and will include the tasks shown in Figure 4.3, which are set out in more detail later:

- Reviewing the project strategy, control systems, procedures and amending the project handbook, as required.

- Amplifying the design brief as necessary during design development.

- In conjunction with the project team, prepare updates to the master development schedule for the detailed design and production information stage, defining tasks and allocating responsibilities.

- Updating the schedule to establish timely flow of information from the design team for

- cost checking

- client's approval

- tender preparations

- construction processes

- Coordinating the activities of the client and the project team in the management of the production of the design information.

- Formulating, in collaboration with the consultants, recommendations to the client/owner in respect of the quality control system, including:

  ▪ on-site and off-site inspection of work for compliance with specifications, and testing of materials and workmanship

  ▪ performance testing and the criteria to be used

  ▪ updating the schedule to incorporate requirements for samples and mock-ups, their updating and monitoring progress of approvals; copy of the schedule is to be included in the relevant monthly reports.

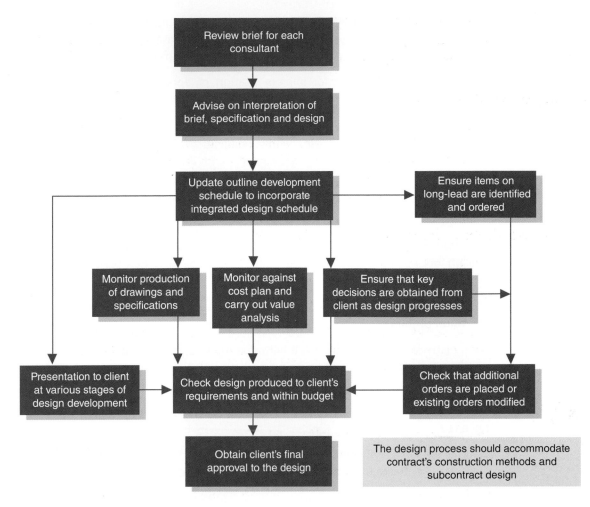

Figure 4.3 Coordination of design work up to design freeze.

- Listing the key criteria in terms of performance benchmarking that in all areas of design make clear how the design will be judged, that is, air changes or faults with current facilities.

- Monitoring the emerging technical design against the risk register, cost plan and the master development schedule.

- Liaising with the client and project team and the local authority and utility companies and other statutory bodies to obtain permissions and approvals.

- Evaluating changes in the client's requirements for cost and time implications and incorporating approved items into the design process.

- Reviewing progress, updating the master development schedule and providing regular reports incorporating information relating to:

  - project status

  - intervening events and their effects

  - cost against budget/cost plan, together with reconciliation statement

  - forecast of total cost and date of completion

  - risk register and contingency planning

  - mitigation and recovery plan

- Obtaining the client's approval to the detailed design and production information phase.

- Coordinate and liaise with the CDM coordinator to ensure arrangements for health and safety of planning and design work to enable satisfaction of the risk management hierarchy (Eliminate – Reduce – Inform – Communicate).

- Initiating arrangements for implementation of approved design and production information, to ensure that contractors' reasonable information requirements are fulfilled.

Value management

Value management, as outlined in the previous section, usually incorporates a series of workshops, interviews and reviews, through which the project requirements can be evaluated against the means of achieving them.

The approach taken on value management will vary depending on the procurement route chosen, but generally it is a technique which encourages to view projects on whole life-cycle basis.

The application of value management, as a formal technique, will largely depend on the value of the project and the level of risk involve.

For a high-risk, high-value project, a full value management procedure is almost always justified. For high-risk, low-value projects or low-risk, high-value projects, desk top value management studies will often suffice, provided there is input from all stakeholders.

For low-risk and low-value projects, a value management exercise is not often necessary.[1]

Value engineering (VE), which forms a key element of value management and requires input from contractors and specialists, has been detailed at a later part of this section.

Contract award

Following the tender procedure, once the entity with the best bid is selected, the contract has to be formally awarded to the successful bidder which also confirms acceptance of the bid and the commitment to budget expenditure. Prior to the award of the contract, typically the funding must be secured in entirety and located within the control of the client.

It is also good practice to write to all unsuccessful bidders advising them the outcome of the process and providing some feedback where appropriate.

Public sector entities will have their own specific processes and procedures governing the Contract award and where the EU Directives apply, there are specific processes and notifications which have to be undertaken for the Contract award process.

Pre-start meeting

The pre-start meeting with contractors and consultants (project team) is held to establish proper working arrangements, roles and responsibilities, lines of communication and agree procedures to be followed throughout the contract (project on-site). If bonds are required they must be provided before possession of site is granted. The principal contractor's construction phase health and safety plan must be in place before work starts on the site.

---

[1]   Value & Risk Management, (Dallas, 2006) – a CIOB publication.

Agenda items at pre-start meeting

Introduction

- Introduce the representatives who will regularly attend progress meetings and clarify their roles and responsibilities. The client, contractor and consultants may wish to introduce themselves.

- Briefly describe the project and its priorities and objectives, and any separate contract that may be relevant (preliminary, client's own contractors, etc.).

- Indicate any specialists appointed by the client, for example, for quality control, commissioning, for this contract.

Contract

- Describe the position with regard to preparation and signature of documents.

- Handover any outstanding production information, including nomination instructions, variation instructions. Review situation for issuing other important information.

- Request that insurance documents be available for inspection immediately, remind the contractor to check specialist subcontractors' indemnities. Check if further instructions are needed for special cover.

- Confirm the existence, status and use of the information release schedule, if used. Establish a procedure for agreeing adjustments to the schedule should they be necessary.

Contractors' matters

- Check that the contractors' working schedule is in the form required and that it satisfactorily accommodates the specialist subcontractors. It must:
  - contain adequate separate work elements to measure their progress and integration with services installations
  - allocate specific dates for specialist subcontract works, including supply of information, site operations, testing and commissioning
  - accommodate public utilities, etc.

- Agree a procedure for the contractor to inform the architect of information required in addition to any shown on the information release schedule. This is likely to involve the contractor's schedule of information required, which must relate to their working schedule and must be kept up to date and regularly reviewed. It should include information, data, drawings, etc., to be supplied by the contractor/specialist subcontractors to the design team.

- Review in detail the particular provisions in the contract concerning site access, organisation, facilities, restrictions, services, etc., to ensure that no queries remain outstanding. Ensure that the contractor has a copy of any conditions placed on the client in respect of the planning consent. Also provide the contractor with legal drawings showing the curtilage of the site ownership.

- Quality control is the contractor's responsibility. Remind the contractor of the contractual duty to supervise standards and quality of work during the execution of the works.

- Determine the information that will require to be provided by the contractor to meet the sustainability strategy such as timber certificates and reports on waste management.

- Numerous other matters may need special coverage, for example,

- check whether immediate action may be needed by the contractor over specialist subcontractors and suppliers

- emphasise that drawings, data, etc., received from contractor or specialist subcontractors which are yet to be approved will remain the responsibility of the originator until approval

- review outstanding requirements for information to or from the contractor in connection with specialist works

- clarify that the contractor is responsible for coordinating performance of specialist works and for their workmanship and materials, and for coordinating site dimensions and tolerances

- The contractor must also provide competent testing and commissioning of services as set out in the contract documents, and should be reminded that the time allocated for commissioning is not a contingency period for the main contract works.

- The contractor must obtain written consent before subletting any work.

Resident engineer/clerk of works' matters

- Clarify that inspections are periodic visits to meet the contractor's supervisory staff, plus spot visits.

- Explain the supportive nature of the various roles and the need for cooperation to enable the clerk of works and resident engineer to carry out their duties.

- Remind the contractor that the resident staff must be provided with adequate facilities and access, together with information about site staff, equipment and operations.

- Confirm procedures for checking quality control, for example, through:

  - certificates, vouchers, etc., as required

  - sample material to be submitted

  - samples of workmanship to be submitted prior to work commencing

  - test procedures set out in the bills of quantities

  - adequate protection and storage

  - visits to suppliers' and manufacturers' works

Consultants' matters

- Emphasis that consultants will liaise with specialist subcontractors only through the contractor. Instructions are to be issued only by the architect/contract administrator. The contractor is responsible for managing and coordinating specialist subcontractors.

- Establish working arrangements for specialists' drawings and data for evaluation (especially services) to suitable timetables. Aim to agree procedures which will speed up the process; this sector of work frequently causes serious delay or disruption.

Quantity surveyor's matters

- Agree procedures for valuations; these may have to meet particular dates set by the client to ensure that certificates can be honoured.

Clarify:

- the process and procedures to deal with foreseeable and unforeseeable changes

- tax procedure concerning VAT and 'contractor' status

- status of precedence among the construction issue specifications, drawings and bills of quantities if appropriate

Communications and procedures

- The supply and flow of information will depend on the updated working schedule and will proceed smoothly if:
    - there is regular monitoring of the contractor's working schedule
    - requests for further information are made specifically in writing not by telephone
    - the design team responds quickly to queries
    - technical queries are raised with the clerk of works (if appointed) in the first instance
    - policy queries are directed to the architect/contract administrator
    - discrepancies are referred to the architect/contract administrator for resolution
- On receiving instructions, the contractor should check for discrepancies with existing documents; check that documents being used are current.
- Information to or from specialist subcontractors or suppliers must be via the contractor.
- All information issued by the design team should be via the appropriate forms, certificates, notifications, etc. The contractor should be encouraged to use standard formats and classifications.
- All forms must show the distribution intended; agree numbers of copies of drawings and instructions required by all recipients.
- Clarify that no instructions from the client or consultants have any contractual significance and should not be acted on by the contractor or any subcontractor but should immediately be referred to the contract administrator for decision; only written instructions from the contract administrator are to be actioned under the contract and all oral instructions must be confirmed in writing. Explain the relevant procedure under the contract. The contractor should promptly notify the contract administrator of any written confirmation outstanding.
- Procedures for notices, application or claims of any kind are to be strictly in accordance with the terms of the contract; all such events should be raised immediately the relevant conditions occur or become evident.
- It is advisable that a communication plan is agreed in advance so as to provide a clear direction to all parties involved, particularly in complex projects with multiple stakeholders.

Meetings

Review format, procedures, timing, participants and objectives of the next stage:

- meetings, that is, site (progress) meetings, policy/principal's meetings and contractor's production information meetings
- site inspections

See Table 4.1 for a specimen agenda for a pre-start meeting.

Contractual arrangements

The project manager has to ensure that all statutory and contractual formalities are in place prior to allowing work to start on site. It may mean that the project manager has to ensure that others have given the relevant notice and, if appropriate, received

Table 4.1 Specimen agenda for pre-start meeting

1.	**Introductions**

Appointments, personal
Roles and responsibilities
Project description

2.	**Contract**

Priorities
Handover of production information
Commencement and completion dates
Insurances
Bonds (if applicable)
Standards and quality

3.	**Contractors' matters**

Possession
Schedule
Health and safety files and plan
Site organisations, facilities and planning
Security and protection
Site restrictions
Contractor's quality control policy and procedures
Subcontractors and suppliers
Statutory undertakers
Overhead and underground services
Temporary services
Project information dissemination

4.	**Resident engineer/architect/clerk of works/lead consultant/design coordinator's matters**

Roles and duties
Facilities
Liaison
Instructions

5.	**Consultants' matters**

Structural
Mechanical
Electrical
Others

6.	**Quantity surveyor's matters**

Adjustments to tender figures
Valuation procedures
Remeasurement if appropriate
VAT

7.	**Statutory utilities**

Gas
Water
Electricity
Drainage
Telecommunication

8.	**Communications and procedures**

Information requirements
Distribution of information

(Continued)

Table 4.1 *(continued)*

Valid instructions
Lines of communication
Dealing with queries
Building control inspections
Notices to adjoining owners and occupiers
9. Meetings
Frequency and proceedings
Status of minutes
Distribution of minutes

the relevant approval. A log will help to keep track of notices and approvals together with the owner of the task.

These are likely to include:

- planning consent with conditions to be discharged

- third-party agreements (such as landlord's approval, party wall and rights of light)

- CDM notification

- insurances (such as professional indemnity, employer's liability, project insurance and third party)

- notice to start work under the Building Regulations

- Fire Regulation compliance

- performance bonds

On completion various completion certificates are required, these should be specified in the particular specification, and would include:

- Fire Regulation compliance

- electrical completion certificate

- test certificates both manufacturing and installation

- lifting beams tests and marking

- Building Regulation compliance

- pressure vessel and boiler certificates

For special buildings or processes, for example, nuclear power projects, pharmaceutical plants, oil and gas facilities and rail infrastructure, particular licences and certificates may be required. If there is any doubt, ask the design team for their advice, then manage the process.

Establish site

Once the design has been finalised and the contracts signed, the project is ready to go on site. It is imperative that the site set-up process is carried out and completed in the most efficient manner prior to the start of the main construction works. The issues that the project manager must be aware of and monitor with the contractor at this stage are not only practical and physical operations but also administrative plans and procedures agreed by the parties. The areas where the project manager is to agree and monitor site set-up are:

- Site boundaries clearly identified with the contractor.

- Establish the contractor's proposal for security.

- Establish the contractor's proposal for emergency plans in case of fire or any serious incidents.

- Establish the contractor's proposal of site accommodation; specifically the suitability of welfare facilities.

- Carry out a survey of existing conditions of the site and the adjacent properties. Record any relevant issues.

- Establish with the contractor the administrative procedures such as request for information (RFIs), confirmation of verbal instructions (CVIs), daily returns, daily diaries, faxes, e-mail facility, drawing issues, etc. This activity is one of the most important action to do as it will set out the communication route between all the participants throughout the project. The findings and agreements with the contractor should be recorded by the project manager and distributed to all professionals involved.

- Ensure that the contractor is aware of, and is attending to, any issues that may be present due to neighbours being close to the site including the terms of any party wall awards or rights of light issues.

- Ensure that the contractor has clearly identified the health and safety risks that exist on the site.

- Ensure all signage is displayed correctly.

The above issues are to be agreed with the contractor. The project manager cannot dictate how the contractor is to set up the site. The project manager's role must be advisory and thus monitor that the correct actions as agreed are being implemented.

Control and monitoring systems

It is the project manager's prime duty to make sure that all necessary control and monitoring systems are properly set up and implemented by the contractor.

The project manager should endeavour that these systems produce the most appropriate information and reports, on a regular and timely basis, so that they can be used to monitor and manage the project to its successful conclusion.

By carrying out audits and checks of the systems, the project manager must be fully satisfied with the accuracy of the information produced and that it does indeed indicate the 'real' position at any point in time and, where appropriate, accurately forecast the final position for the project.

Contractor control and management systems will generally be (but not limited to) the following:

- quality management system

- schedule management system

- quality control system

- cost monitoring and management system

- health, safety and welfare system

- environmental management system including waste management

- IT and communication

- document management and sharing system

It is of absolute importance that the project manager fully understands the relevance of the information being produced from these systems. The project manager must proactively use this information to manage the contractor and the project team through the regular management meetings. The aim is not only to understand where the project is and where it is ultimately going, but also to identify any potential problem areas at such a sufficiently early stage so that any rectification procedures and/or mitigation measures can be taken to ensure best delivery of the project.

Contractor's working schedule

The project manager has a duty to the client to monitor the performance of the contractor. In order to adequately carry this out, the project manager needs to ensure the contractor has prepared a construction schedule (working schedule) in sufficient detail to enable the construction works to be closely monitored.

The project manager needs to receive and review the contractor's working schedule prior to the commencement of the works in order to:

- check it complies with the client's time requirements
- check it acknowledges any restraints imposed on the construction of the works
- ensure that the level of detail is appropriate for the illustrating the progress of the works
- ensure it suitable for monitoring the progress of the works
- confirm the sequencing and logic of the schedule

The working schedule must incorporate an information requirement schedule so that it realistically informs the project manager when outstanding design information is required in order for the contractor to achieve the schedule dates. Regular reports recording the progress achieved against the schedule must be received from the contractor, and a progress status agreed with the contractor.

Any rescheduling of the works necessary to recover delay situations need to be received, reviewed and agreed. In addition to the detailed analysis of progress, the project manager should examine high-level progress trends to obtain an overall view of project status. This can involve graphically comparing accumulative planned progress against actual achieved. Typically a contract will require that the contractor's tender schedule becomes part of the contract. This schedule does not normally go into a great deal of detail, as timescales, dependencies and interfaces may yet to be agreed with subcontractors to the main contractor. The project manager will need to obtain a working schedule detailing all specific sections of work. Rescheduling required as a result of changes or slippages will be required to show how time can be recovered or the effect on the completion dates.

It is the duty of the project manager to not only to monitor the contractor's progress, but also to monitor any work being undertaken by other advisors, suppliers or companies that have an independent input into the completion of the project. These should all be monitored against the development schedule with its own milestones and targets. The project manager is managing the overall project for the client and its successful delivery.

Value engineering (related to construction methods)

Value engineering (VE) is an exercise that most of the project team undertakes as the project develops, by selecting the most cost-effective solution. However, VE is about taking a wider view and looking at the selection of materials, plant, equipment and processes to see if a more cost-effective solution exists that will achieve the same project objectives.

Table 4.2 Value engineering job plan

Information
Function analysis
Speculation
Evaluation
Development
Recommendation
Implementation

Table 4.3 Result accelerators

Avoid generalities
Get all available costs
Use information from best source
Blast, create and refine
Be creative
Identify and overcome roadblocks
Use industry experts
Price key tolerances
Use standard products
Use (and pay for) specialist advice
Use specialist processes

VE should start at project inception where benefits can be greatest; however, the contractor may have significant contributions to be made as long as the changes required to the contract do not affect the timescales, completion dates and incur additional costs that outweigh the savings on offer. There is, however, still a place for VE, especially at the start of construction. The application of the job plan (see Table 4.2) remains consistent, but the detail available is obviously more than during the design and pre-design stages. The 'results accelerators' still act as useful guides to VE at the construction stage (see Table 4.3). In all of this, it is most important to remember the relationship between cost and value: value is function divided by cost. Concentrating on the function of the project or product will avoid mere cost cutting.

The project manager must take a proactive role in both giving direction and leadership in the VE process, but must above all ensure that time and effort is not wasted and does not have a detrimental effect on the progress of the project. An example of a VM framework has been included in Briefing Note 3.02.

Management of the supply chain

The contractor has overall responsibility for the management of the supply chain to meet the contractual obligations. The project manager has the duty to ensure that this chain is being effectively managed so as to avoid any potential delay, unnecessary cost implications or any other adverse effect on the delivery of the project. This is an important issue as it is so very often the case that problems further down the contractual chain can be responsible for long delays and/or major disputes right back up through the whole chain. They can potentially result in the deterioration of relationships with the contractor and have a knock-on effect with not only the contractor's performance but also the performance of the project team as a whole.

Duties and responsibilities should include:

- Receive and understand the details of the contractor's supply chain and the controls to manage it.

- Establish key members and linkages within the chain.

- Receive and interrogate reports from the contractor of the ongoing progress, including any reports from procurement managers and expeditors.

- Implement a regular monitoring system to check the progress of key suppliers or subcontractors (against the contractor's delivery schedule) so that there are timely warning signals of any potential delays or failures that could have an adverse effect on the progress and financial stability of the project.

- Agree with the contractor any appropriate remedial action that may be needed to rectify any problem areas.

Risk management

The risk register (see Briefing Note 3.03) is a document that should be prepared at the earliest stages of the project, identifying potential risks throughout the project. This register should be reviewed and updated according to circumstances, and stages of the contract. At the construction stage the risk register should be reviewed to include any new construction risks.

In addition to monitoring those construction-related risks previously identified in the project-wide risk register, the project manager should require the contractor to instigate and maintain a risk management system for those risks likely to impact on the actual construction works. The project manager should also require the contractor to:

- establish a fully detailed listing of construction risks

- determine the likely probability and impact of each risk

- review the risks with the project team

- prepare method statements and action plans demonstrating how risks will be mitigated or managed out

- identify, and inform, the person responsible for managing each risk

- prepare contingency plans for any key risks having a significant impact

- regularly review and report on the status of risks

Payments

The traditional practice in the construction industry is for the contractor to be paid monthly during the execution of the works. The value of these payments is determined by agreement between the respective Quantity Surveyors of the employer and the Contractor. An advance based on measurement is a system of payment which requires detailed and time consuming management. It does not reward achievement nor does it distinguish between the inefficient and efficient contractor. It is questionable whether it delivers best value for money.

It is often said that cash is king and in construction contracting, cash is the contractors' (and subcontractors') primary concern. Over the years, contractors have come up with innovative ways of enhancing cash flow. Some of these ways have been found in more efficient management processes and information systems by which contractors minimise outstanding balances owed by clients. Some have been found through pricing policies (e.g. unbalancing and front-end loading) or somewhat unfair procedures such as over-measurement and delaying payments to subcontractors and suppliers. The extent to which cash flow influences planning of construction activities on site is however an unknown element and an issue that has not attracted much interest yet. One possible reason for this perhaps is the traditional mechanism of payment itself.

Construction programmes although will undoubtedly influence clients' cash flow, they have very limited effect on contractors' net cash flow unless the payment mechanisms are based on programme milestones or activities.

Contractors get paid (typically monthly) for the work they manage and in turn distribute most of these payments to subcontractors and suppliers in accordance with their contractual arrangements (leaving them with the balance which is usually overhead and profit minus retentions).

The recent changes in legislation are aimed to simplify the payment mechanisms in the construction supply chain and to completely banish the 'pay when paid' scenario between the main contractors and the subcontractors and suppliers.

Benchmarking

In certain circumstances, particularly when framework or partnering agreements are in place, it may be appropriate to employ benchmarking of a contractor's performance against the best industry practice. A major difficulty of benchmarking in construction is locating base data that allows for meaningful comparisons. Since 1998, as part of the annual production of statistical information gathered from contractors, the British government collected measures of key performance indicators. These provide the widest sourced comparators currently available to benchmark individual companies against the industry average levels of performance.

A number of construction clients commission their own research to assemble from other similar organisations meaningful performance data that allows them to carry out benchmarking of the companies they use.

Benchmarking is closely associated with the concept of continuous improvement, and a company's performance can be monitored over time to confirm that measures introduced to lead to improvement are effective.

Change and variation control

The project manager should carry out the following tasks to control variations:

- Monitoring and controlling variations which result from changes to the project brief to be avoided whenever possible (Figure 4.4) or design/schedule modification (e.g. client's request, architect's or site instructions) must follow a procedure which:

  - identifies all consequences of the variation involved
  - takes account of the relevant contractual provisions
  - defines a cost limit, above which the client must be consulted and, similarly, when specifications or completion dates are affected
  - authorises all variations only through the appropriate change procedure

- Identifying, in consultation with the project team, actual or potential problems and providing solutions which are within the time and cost limits and do not compromise the client's requirements, with whom solutions are discussed and approval obtained.

- Checking the receipt of scheduled and/or ad hoc reports, information and progress data from project team members.

The main effect on the reduction of claims or variations is to ensure the brief is clearly defined, and the contract documents and drawings accurately and completely reflect the detail (Table 4.4).

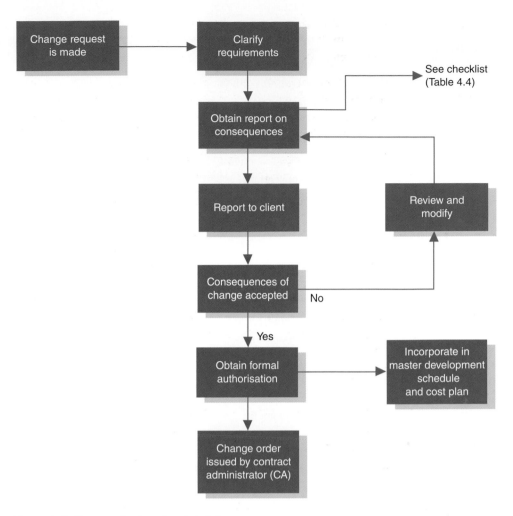

Figure 4.4 Changes in the client's brief.

Managing change control at design development stage is far more effective than managing the process when construction is in progress. Circumstance-driven changes, mistakes or unknowns have to be effectively managed on the basis that in many instances time is more expensive than the material change. Some form of authorisation may need to be agreed (typically delegated financial limits) so that instructions can be given without having to refer every change back to the client for approval.

The project manager will need to maintain a register of changes and variations, cross-referenced to contractor's RFI notices, and possibly contract claims. The register should include budget costs and final costs for reporting to the client on a regular basis.

Accurate, detailed daily diaries will need to be kept, complete with plant, labour and material deliveries so that consequential costs can be identified.

In dealing with the effects and costs of variations, the project manager in conjunction with the contract administrator will need, where possible, to agree costs before issuing an instruction. It is also wise to agree, again where possible if work will be undertaken with no overall effect on the schedule. It is vital to carefully record events and situation at the time.

Procedurally, the project manager must inform both the design consultants and the main contractor that all variation instructions must be in the correct written form and *must* only be issued via the contract administrator unless the project manager

Table 4.4 Changes in the client's brief: checklist

Activity	Action by
1. Request for change received from client	Project manager
2. Client's need clarified and documented	Project manager/contract administrator (CA)
3. Details conveyed to project team	Project manager/CA
4. Review of technical and health and safety implications	Consultants and project manager
5. Assessment of programme implications	Planning support staff and project manager
6. Evaluation/calculation of cost implications	Quantity surveyor
7. Review engineering services commissioning	Commissioning manager
8. Preparation of report on effect of change	Project manager in consultation with
9. Reporting to client	Project manager
10. Consequences accepted/not accepted by client	Project manager/CA
11. Non-acceptance: further review/considerations as per items 4–6 and action items 7 and 8	Project manager assisted by consultants
12. Further reporting to and negotiation of final outcome with client	Project manager assisted by consultants
13. Agreement reached and formal authorisation obtained	Project manager/CA
14. Incorporation into master development schedule and cost plan (budget)	Project manager and quantity surveyor
15. Change order issued (see Briefing Note 3.13)	CA

is the appointed contract administrator to the main contractor. To avoid unnecessary complications in agreeing valuations and accounts, it is imperative that the variation instructions are issue from one source. Design consultants must raise (in writing) RFI to the project manager who will in turn issue the instructions to the contractor. *All* variations *must* have an instruction (in writing) against them in order to be valued.

Dispute resolution

The Woolf reforms (1999)[2] recommended that litigation should be seen as a last resort and alternative dispute resolution such as mediation and other methods of resolving disputes are actively encouraged; parties who do not participate in mediation must justify their position to a judge.

Adjudication may be used to resolve disputes using the provisions of the Housing Grants, Construction and Regeneration Act, 1996 and subsequent amendments (see Briefing Note 4.03).

Although it is hoped that the non-adversarial approach and the increasing choice of alternative procurement options and partnering will lead to a reduction in disputes, nevertheless, the project manager should make every effort to pre-empt any dispute that may arise and endeavour to mitigate and resolve the problem.

---

[2] Lodr Woolf, July 1996, Access to Justice Final Report, webarchive.nationalarchives.gov.uk/+/http:/www.dca.gov.uk/civil/final/index.htm (accessed April 2014).

Arbitration is a procedure that may be used following adjudication and before initiating litigation. An outline of procedures to be applied in resolving contractual disputes is given in Briefing Note 4.02.

BIM strategy

The project BIM execution protocol will establish the progression of the level of development or detail specification for the project BIM and requirements for associated data sets. The level of detail and maturity of information will relate to the relative design stage. The Level of Development (LOD) specification will define information required at every stage from all the designers and any other stakeholders inputting to the model, including supply chain with a design responsibility. It will also define permitted uses of the model at each stage for say, estimating, programming and so on.

The integrated design schedule will define the key milestones, activities and outputs linked to the LOD spec stages, so that at each stage it is clear what level of detail the design team need to achieve in the model. These are illustrated in the RIBA BIM Overlay, stages 1–7.[3]

---

[3] BIM Overlay to the RIBA Outline Plan of Work, 2012.

Briefing Note 4.01   Regular report to client

Notes for guidance on contents

Executive Summary

The purpose of the executive summary is to give the client a snapshot of the Project on a particular date which can be absorbed in a few minutes. It should contain short precise statements on the following:

- Significant events that have been achieved.
- Significant events that have not been achieved and action being taken.
- Significant events in the near future, particularly where they require specific action.
- Progress against the master, design and construction schedules.
- Financial status of the project.

Contractual arrangements (including legal agreements)

Each project requires the client to enter into a number of legal agreements with parties such as local authorities, funding institutions, purchasers, tenants, consultants and contractors. The report should be subdivided to identify each particular agreement and to provide details of requirements and progress made against the original project master schedule. The following are indications of possible legal agreements that may be required on a project:
- joint development agreement
- land purchase agreement
- funding agreement
- purchase agreement
- tenant/lease agreement
- consultants' appointments
- Town and Country Planning Acts: sections in force at the time, for example,
  - planning gain
  - highways agreement
  - planning notices
  - land adoption agreement
  - public utilities diversion contracts

Client's brief and requirements

This provides a 'status' report on how the client's brief and requirements are progressing. The report should identify any requirements which need clarification or amplification and also those which are still to be defined by the client.

Client change requests

Client-orientated changes should be listed under status (being considered, in progress, completed), cost and schedule implications. The objective is to make the client fully aware of the impact and progress of any change.

(Continued)

Planning building regulations and fire officer consents	This section will be subdivided into the various consents required on a specific project. Each section should highlight progress made, problems possible solutions and action required or in progress. The following are examples of possible consents: • planning – outline • planning – detailed, including conditions • Building Regulations • means of escape • English heritage/historic buildings • fire officer • public health • environmental health • party-wall awards
Public utilities	Each separate utility should be dealt with in terms of commitment, progress, completion and any agreements, way-leaves as appropriate.
Design reports – summaries	The design team and consultants should prepare reports on progress problems and solutions which will form the appendices and must include marked-up design schedules and 'issue of information' schedules. The design report, however, should be distilled into an 'impact-making' synopsis and agreed as a fair representation by each member concerned.
Health and safety	Report on the preparation of the CDM health and safety plan and the health and safety file.
Project master schedule	Updated schedules should form an appendix to the report, specifying progress made. A short commentary on any noteworthy aspects should be made under this section.
Tendering report	This is a status report on events leading up to the acceptance of tenders. It should show clearly how the various stages are progressing against the action plan.
Construction report summary	This report is prepared in a similar way to that outlined for design reports (above).
Construction schedule	The updated schedule should form an appendix to the report, highlighting progress made and showing where delays are occurring or are anticipated. A short commentary on any important items should be given in the report under this section.
Financial report	A fully detailed financial report should form one of the appendices. It should provide a condensed overview (say two to three pages) giving the financial status and cash flow of the project. This report will embrace the information provided by the quantity surveyor and also call for the project manager to provide an overall financial view, highlighting any specific matters of interest to the client.
Appendices	These will include full reports and schedule updates as outlined in the previous sections. Other reports, possibly of a specialist nature, may also be included. Should the report be presented at a formal meeting, then the minutes of previous meetings should be included in the appendices.

Briefing Note 4.02 Dispute resolution methods

The construction industry has been a fertile breeding ground for disputes. They cannot be avoided entirely and it would be foolish to suggest that they could. Among other things, there may be design faults, there can be defective work or materials, the cost of variations may cause dismay, money can be wrongfully withheld, loss and expense for delay and prolongation, or extensions of time to defend against liquidated damages for late completion may be claimed.

On the other hand, the high cost of energy-sapping defended litigation can often be avoided by sensible planning your dispute resolution procedures before contract as well as by the proactive management of the process of resolution once a dispute has arisen.

Mediation, conciliation, expert determination, adjudication, arbitration and, of course litigation are all possibilities to be considered. Two of these – mediation and conciliation – are often referred to as 'ADR' an acronym that means 'alternative dispute resolution'. That in itself does not mean much without recognising to what it is an alternative. The essential difference between orthodox dispute resolution and ADR is that in ADR the parties make their own settlement agreement, which is only binding so long as they want it to be, and in orthodox dispute resolution the decision is made for them by a third party and it is final and binding upon them. There is a grey area in all this and that is in expert determination and adjudication in which the decision can be final and binding, or it can be final and binding unless disputed in another forum, or it could be non-binding depending upon how (and under what law) it is structured.

Apart from reference to the courts by litigation (which in every common law country is a unilateral act, open to anyone who think they have had a right infringed) all the other methods of dispute resolution require an agreement. Naturally, it is easier to agree a method of resolving a dispute before it has arisen rather than after. However, irrespective of whether there is an agreement in place, it is always open to either party to suggest an alternative means of dispute resolution that will save both parties time, cost and frustration, and to enter into an agreement for that, at any time.

Non-binding

In non-binding processes, the dispute resolver helps the parties to agree their differences. These are entirely private processes, conducted without prejudice to the rights of either party, and there is nothing stopping either party from shifting its ground during the process. Indeed, if it is to be successful, it is essential that they do. If they do not succeed in reaching a settlement, there is nothing to prevent either party from dealing with the same dispute through another forum at a later date and nothing that has been discussed in the ADR process may be used in evidence can usually be used elsewhere. The dispute resolver will agree with both parties a procedure; he will read the parties' respective position statements and any documents provided in support. He will consult with the parties privately, and with both together.

Although essentially a non-binding process, it is always open to both parties to agree that the final settlement should be binding. The parties agree to share the costs of the dispute resolver and to pay their own. This is an excellent method of dealing with disputes because it encourages the parties to talk to each other, if successful it helps to preserve working relationships and, even if unsuccessful, it helps the parties to focus on the real matters in which they are in dispute. In many contracts, ADR is required at some stage and in England, court ordered ADR forms a part of the Civil Procedure Rules of the courts.

Mediation

Without express permission, the mediator will never disclose what has been said to him by either party to the other. A mediator does not have to have a detailed understanding of the facts or the law of the matters in dispute but it often helps. He will not advise the parties of their rights nor generally will he advise the parties of the strength of their case but he will help each to see the weaknesses of their own and the strengths of their opponent's position. In doing so, he will draw them closer together with a view to executing an agreement to settle their differences. In general, mediation can be completed in two to three days. In very large cases with many issues, it might take a week or more but that is unusual.

Conciliation

Conciliation as a similar process to mediation but the conciliator takes a more active role in the settlement of the dispute than does the mediator. A conciliator necessarily has to have to a detailed understanding of the facts and law of the matters in dispute. The conciliator will express an opinion on the relative merits of the parties' respective cases. He will try to persuade them of his views and, in doing so, will attempt to guide the parties into an agreement compatible with the parties' rights under the contract. Conciliation can be expected to be a little shorter than mediation simply because the conciliator is able to focus the parties' attention on the issues and drive the process in a way that is unavailable to a mediator. In general, conciliation can be completed in one or two days. As with mediation, in very large cases with many issues it might take a week or more but that is unusual.

Non-binding or final and binding

Unlike ADR in which the parties make their own decision, the essence of these decision-making processes is that a third party is introduced to make the decision for them. Because the process is consensual, it is always a private process. However, depending upon the rules of engagement agreed between the parties, the information that becomes available may not be privileged and the decision made may not be binding on the parties, leaving them free to revisit the dispute in another forum. The parties are free to agree who should pay the dispute resolver's costs and how the party's costs should be dealt with, although it is usual for each side to pay their own costs.

Expert determination

Expert determination is quite different from any other method of dispute resolution. In this forum, the expert is appointed for his knowledge and understanding of the particular issues in dispute in the field in which he is an acknowledged expert. The expert will agree with both parties a procedure; he will read the parties' respective position statements and any documents provided in support. There is usually no provision for the parties to change their position or amend their case during the process. He will consult with the parties privately and may consult with both together, but he is under no obligation to do so unless it is made a term of his appointment. He is given the role of investigator. He is required to find the facts and law in relation to the issues in dispute, to make his own inquiries, tests and calculations and to form his own opinion and decide upon the merits of the parties' position. Depending upon the issues, expert determination can involve much research and a hearing and can take anything from a week to several months.

Adjudication

In England and Wales and in several Commonwealth countries, adjudication has recently been given statutory authority. Under the law of those countries that adopt this process, it is generally the rule that either party to a specified type of construction contract has the right at any time to submit any dispute or difference to the adjudication of a third party. However, even where the statutory right is limited to particular types of contract, there is nothing stopping the parties from agreeing by contract to follow the same process in regard to contracts which are outside the Act and that is common.

Adjudicators are often appointed for their knowledge and experience of the type of matters in dispute, although it is not essential. Although the idea of adjudication is that there should be a decision, in the event that the parties do not like the result there is nothing to prevent them from running the case again in another forum; the rule of *res judicata* does not apply to adjudication. The adjudicator will agree with both parties a procedure; he will read the parties' respective referrals and any documents provided in support. He may also require a hearing and will often conduct conference calls with the parties.

The adjudicator's decision is binding until either party decides to refer the same dispute to arbitration or litigation, in which case the decision is binding until an award or judgement is handed down. When the legislation was first enacted in England in 2000, the adjudicator was empowered to make his own inquiries of the facts and law. It was thought that he might act pretty much like an architect or engineer under a construction contract and that few parties would take the adjudicator's decision as final and binding, so it was not initially thought necessary for the adjudicator to act within the rules of natural justice.

Five years on and several hundred enforcement cases later, it became clear that parties who have been unhappy with the outcome have sought to overturn the decision on the basis of the adjudicator's misconduct rather than have the case rerun in arbitration or litigation. As a result, the courts have imposed upon adjudicators the obligation to act within the rules of natural justice. They must hear both sides. The parties must have an equal opportunity to make their own case and to respond to the case against them, although they may not alter or amend their submissions. This is a tall order in the limited time available to make the decision. They must be impartial but they do not have to be independent. They may only inquire into the facts and the law of the cases that are put to their decision. They may not go outside the parameters of the parties' submissions to make good any deficiencies.

Unless the referring party agrees to extend the period for the decision by up to 14 days, or both parties agree to extend the period the decision beyond that the dispute resolution process must be conducted and the decision given within 28 days of referral. The adjudicator has no power to order discovery or to take evidence on oath unless the parties give it to him by agreement and if either party request it, he must give reasons for his decision...and it all seems to work very satisfactorily.

Final and binding

In the sense that in the following tribunals the facts once found cannot be re-opened by any court, the matters are *res judicata*. Appeal on a point of law is always available from a domestic arbitral tribunal to the court and from a lower court to a higher court. However, statute has tended to limit the right of appeal from an arbitrator's award in other than a point of law of public importance in order to give the parties a greater sense of finality.

Arbitration

An arbitration agreement is written into all standard forms a building and civil engineering contract. It is a private process and nobody is permitted to know of the matters in dispute or the decision unless the parties agree otherwise. The arbitrator's decision is final and binding and can be enforced in many countries of the world by virtue of the New York Convention. Arbitrators, like judges, must be independent and impartial. They must scrupulously follow the law of the contract and the rules of natural justice to provide a speedy and efficient decision on all the issues submitted to jurisdiction. The arbitrator may not go outside that limitation to decide things that were not part of the reference.

Subject to the arbitration agreement, the parties may adopt specific procedural rules which dictate the powers of the arbitrator or the procedure to be followed. Otherwise, the powers of the arbitrator are set out by statutory instrument. In domestic disputes, it is normal for the reference to be to a single arbitrator, but in international disputes it is more common for each party to appoint their own arbitrator and for the arbitrators to appoint a chairman or umpire, forming a three-man tribunal.

Arbitration can be very time consuming and expensive or it can be quick and cheap depending upon the parties and the case management skills of the arbitrator. There is usually nothing to stop a party from amending its case subject to paying the costs of the other side thrown away. Generally, the arbitrator has the powers of a High Court judge in regard to the taking of evidence on oath, subpoenas for evidence, discovery and so on. He can order a party to pay the costs of interlocutory matters and can determine who should pay his fees and whether the losing party should pay the winning party's costs, in whole or in part, with or without interest and on what basis. The arbitrator must give reasons for his decision if either party requests it.

Litigation

Litigation is the dispute resolution process run by the civil courts of the state. It is free to every individual who has a grievance to resolve. Judges tend not to be technical people, although in some courts they are specifically selected for their technical ability (e.g. the English Technology and Construction Court). On the other hand, judges often have the power to appoint technical assessors or experts to assist them and will almost always do so if the parties request it.

Notwithstanding that the court and the judge are provided by the state, litigation is often a very expensive process. This is often simply because of the complicated rules of procedure, which a reluctant but wily litigant can often exploit to put off the hearing of the case for years, including amending its case from time to time. There are also restrictions on who can appear in the courts on behalf of a litigant. In large cases, the costs can run to many thousands of pounds per day during a hearing, which may take many months or even years before the dispute reaches that stage.

Litigation is a public process (justice must be seen to be done), and the public are encouraged to sit in on the proceedings to hear of the matters in dispute. Judges must give reasons for their decisions, and important decisions are published and recorded in law reports (Pickavance, 2005).

Briefing Note 4.03   Implications of Housing Grants, Construction and Regeneration Act 1996, Amended 2011

The Housing Grants, Construction and Regeneration Act, also known as the 'Construction Act', has been an important part of the law affecting the construction industry since it came into force on 1 May 1998.

Part 8 of the Local Democracy, Economic Development and Construction Act substantially amends the Construction Act. It affects all 'construction contracts' in England, Wales and Scotland. The amendments to the Construction Act came into force in relation to construction contracts entered into on or after 1 October 2011 in England and Wales, and 1 November 2011 in Scotland.

The aims of the amendments are:

- to increase clarity and certainty as to payment in construction contracts

- to introduce a 'fairer' payment regime and improve rights for contractors to suspend their work in non-payment circumstances

- to make adjudication more accessible for the resolution of disputes

Salient points introduced through the amendment include:

- payment notices are crucial as if there is no payment notice the other party can serve a notice of default or rely on its own application for payment

- the paying party will then have to pay whatever has been notified unless a valid notice of the intention to pay less has been served

- payment clauses will need to be redrafted to reflect these changes, and those operating on the construction site must be made aware of the importance of sending the correct notice on time and with the correct information

For details of the current Act, see www.legislation.gov.uk.

Briefing Note 4.04   Typical meetings and their objectives

Steering group/team

- to consider project brief, design concepts, capital budget and programmes

- to approve changes to project brief

- to review project strategies and overall progress towards achieving client's goals

- to approve appointments for consultants and contractors

Project team

- to agree cost plan and report on actual expenditure against agreed plan

- to review tender lists, tenders received and decide on awarding work

- to report on progress on design and construction programmes

- to review and make recommendations for proposed changes to design and costs, including client changes; to approve relevant modifications to project programmes

Design team

- to review, report on and implement all matters related to design and cost

- to determine/review client decisions

- to prepare information/report/advice to project team on (1) appointment of sub/specialist contractors; (2) proposed design and/or cost changes

- to review receipt coordination and processing of subcontractors' design information

- to ensure overall coordination of design and design information

Finance group/team

- to review, monitor and report financial, contractual and procurement aspects to appropriate parties

- to prepare a project cost plan for approval by the client

- to prepare and review regular cost reports and cash flows, including forecasts of additional expenditure

- to review taxation matters

- to monitor the preparation and issue of all tender and contract documentation

- to review cost implications of proposed client and design team changes

Project team (programme/progress meeting)

- to provide effective communication between teams responsible for the various phases of the project

- to monitor progress and report on developments, proposed changes and programme implications

- to review progress against programmes for each stage/section of the project/ works and identify any problems

- to review procurement status

- to review status of information for construction and contractors' subcontractors' requests for information

Project team (site meeting)

Main contractor report tabled monthly to include details on:

- quality control

- progress

- welfare (health, safety, canteen, industrial relations)

- subcontractors

- design and procurement

- information required

- site security

- drawing registers

- reports/reviews (including matters arising at previous meetings) from:

  - architect

  - building services

  - facilities management

  - information technology

  - quantity surveyor

- Statutory undertakings and utilities:

  - telephones

  - gas

  - water

  - electricity

  - drainage

4.04 Briefing Note

- Approvals and consents:
  - planning
  - Building regulations
  - local authority engineer
  - public health department
  - others
- Information:
  - issued by design team (architect's instructions issued and architect's tender activity summary)
  - required from design team
  - required from contractor

5 Construction

Stage checklist

Key processes: Performance monitoring and control

Health, safety and welfare systems

Quality management and control

Key objective: 'Are we constructing what has been designed?'

Key deliverables: Performance management plan

Key resources: Client team

Project manager

Design team

CDM coordinator

Constructor team

Stage process and outcomes

During this stage, the construction of the facilities as defined by the design and contract documentation are completed. This process involves the greatest number of people and organisations, and the greatest expenditure.

The project manager's role during this stage focuses on monitoring the progress of the works, reporting to the client, protecting the client in regard to timescale, cost and quality, and ensuring full compliance with statutory, legal and contractual requirements.

Outcomes:

- progress reports

- contract management and administration

- management of change

- dispute avoidance/resolution (if necessary)

- update Health and Safety Plan and File

- record of construction works

- facilities completed in accordance with design and contract documentation

- as-built drawings

- operating and maintenance manuals

- occupier's handbook

Code of Practice for Project Management for Construction and Development, Fifth Edition. Chartered Institute of Building.
© Chartered Institute of Building 2014. Published 2014 by John Wiley & Sons, Ltd.

Project team duties and responsibilities

Client

Traditionally, the client has had a relatively nominal direct involvement in the construction works; however, as more and more client project teams are being constituted with extensive construction background, the role of the project manager in managing the client's expectations is also expanding. There is now a greater emphasis on the client having more involvement during the construction stage with their primary interests being:

- Ensuring that the build quality is acceptable, taking advice from the project manager where appropriate

- Progress of the works is to schedule and in a logical fashion

- Understanding potential effects of client changes to the construction stage progress

- Managing internal stakeholders in terms of decision making to help with the progress on the site

- Ensuring security, environmental friendly and safe working practices are adopted

- Satisfying themselves that the contractor's performance is in accordance with the contract

- Making sure the obligation to pay all monies certified for payments to consultants and the contractor(s) is being carried out

Project manager

The project manager has a role which is principally that of monitoring the performance of the main contractor and the progress of the works, and involves the following activities (some of which may have been accomplished in the pre-construction stage):

- Ensuring contract documents are prepared and issued to the contractor.

- Ensuring the contracts are signed

- Arranging the handover of the site from the client to the contractor

- Reviewing the contractor's working schedule and method statements

- Ensuring the contractor's resources are adequate and suitable

- Ensuring procedures are in place and being followed

- Ensuring site meetings are held and documented

- Monitoring construction cash flow

- Reviewing progress with the contractor

- Monitoring performance of the contractor

- Ensuring that the construction phase health and safety file is being maintained

- Ensuring design information required by contractor is supplied by consultants

- Establishing control systems for environmental sustainability, time, cost and quality

- Ensuring site inspections are taking place

- Confirm insurance cover on the works

- Managing project cost plan

- Ensuring that the client meets contractual obligations (i.e. payments)

- Reporting to client

- Managing introduction of changes

- Ensuring statutory approvals are being obtained

- Ensuring all relevant legal documents are in place (such as collateral warranties and performance bonds among others)

- Review construction risks

- Establish mechanisms for dealing with any claims

- Monitor for potential problems and resolve before they develop

Design team

The design consultants are responsible for the following:

- Providing production information (i.e. details of building components)

- Commenting and approving working drawings being provided by specialist contractors

- Responding to site queries raised by the contractor

- Inspecting the works to check compliance with the drawings and specification

- Inspecting the works to check an acceptable quality standard has been achieved

Most building contracts refer to a contract administrator, usually the design team leader or the project manager, who is the formal point of contact between the project team and the contractor, and who has a contractual obligation in relation to the issuing of formal instructions to the contractor; these include the following:

- issuing of design information

- issuing of variations

- instructions on standards of work and working methods

- arbitrating on contractual issues

- issuing practical completion certificate

Quantity surveyor

The quantity surveyor has a duty to:

- measure the value of work executed by the main contractor

- agree monthly valuations with the main contractor

- agree the final account with the main contractor

The quantity surveyor has a separate responsibility to the client, usually through the project manager, for reporting on the overall financial aspects of the project.

Contractor

The contractor has several statutory and contractual responsibilities that must be enacted in order to allow the construction of the project to proceed. Depending on the precise form of contract, these responsibilities will vary but will generally include the following:

- Executing the contract agreement between the employer and contractor

- Submitting the requisite health and safety documentation

- Actioning compliance with the requirements of the CDM Regulations of 2007 (see Briefing Note 3.01)

- Implementing the site waste management plan as required under the Site Waste Management Plans Regulations of 2008

- Producing documentary evidence of all insurance policies as required by the contract

- Enacting all parent company guarantees, bonds, warranties, indemnities and third party rights as required by the contract

- Actioning any statutory notices and consents such as planning requirements, hoarding licences, scaffold licence

- Actioning any third party notices, licences and consents such as tower crane over-sailing agreements

- Gaining any necessary consents from the employer such as subletting any part of the works

- Providing the working schedule with all relevant method statements and activity schedules

- Mobilising all necessary labour, subcontractors, materials, equipment and plant in order to commence the construction works in accordance with the contract

Construction manager

A client may decide on a construction management route, directly employing a construction manager as a consultant acting as an agent with expertise in the procurement and supervision of construction and not a principal. In this arrangement, the construction manager's role is the following:

- to determine how the construction works should best be split into packages

- to produce detailed working schedules

- to determine when packages need to be procured

- to manage the procurement process

- to manage the overall site facilities (such as access, storage, welfare, etc.)

- supervise and coordinate the works package contractor's execution of the works

Management contractor

In the managing contracting arrangement, a management contractor acting as a principal would have the additional direct contractual responsibility for the performance of the works package contractors.

Subcontractors and suppliers

Subcontractors have specialist expertise, usually trade related (i.e. mechanical or electrical installations, lift installation, joinery and demolition), for the supply and installation of an element of the total works.

Subcontractors may be either nominated or named by the consultants or selected and appointed directly by the main contractor, known as domestic subcontractors. If nominated, the client carries some risk in respect of the subcontractor's performance.

Suppliers provide certain materials, components or equipment for others to install.

Labour-only subcontractors provide only labour to carry out the installation of materials, components or equipment provided by the main contractor (i.e. carpenters, bricklayers and plasterers).

Due to their specialist knowledge, subcontractors have an increasing design responsibility for the technical design related to their installations (may include fixing details, fabrication details, coordination with other installations).

There is a general obligation on all the project team to ensure the site is safe, although legally this falls to the principal contractor under the CDM Regulations.

Other parties

A large number of other bodies will be involved during the course of the construction works, these include the following:

- building control officer
- highways authority
- environmental health officer
- fire officer
- Health and safety executive
- planning officers
- archaeologists
- trade unions
- landlord's representatives
- funder's representatives
- police

Performance monitoring

Throughout the last few decades, a number of industries, primarily manufacturing, have introduced innovative methods and techniques to shift traditional paradigms in order to improve their performance. This has led to the creation of philosophies such as concurrent engineering/construction, lean/agile production/construction and many others such as JIT (just in time), TQM (total quality management), etc. The main driver behind those philosophies is to optimise an organisation's performance both internally and externally within its respective marketplace. Inevitably, this has led to the 'rethinking' of performance management systems through effective performance measurement.

The construction industry's core business is undertaking projects in generating new buildings or refurbishing existing ones for a variety of clients. Therefore, it is not a surprise to find that traditionally performance measurement in construction is approached in two ways:

(a) in relation to the product as a facility

(b) in relation to the creation of the product

In particular, the latter of the two has been the prime performance assessment (in terms of success or failure) of construction projects.

Although the 'three traditional measures or time, cost and quality' provide an indication as to the success or failure of a project, they do not, in isolation, provide a balanced view of the project's performance. Furthermore, their implementation in construction projects is usually apparent at the end of the project, and therefore they are often classified as 'lagging' rather than 'leading' indicators of performance.

Therefore, the traditional measures of the performance of construction projects are not enough to assess their 'true' performance.

Key performance indicators (KPIs) were devised to generate information on the range of performance being achieved on all construction activity and they generally, in a typical project, these will comprise of:

1. client satisfaction – product
2. client satisfaction – service
3. defects
4. predictability – cost
5. predictability – time
6. profitability
7. productivity
8. safety
9. construction cost
10. construction time

These KPIs are intended for use as benchmarking indicators for the whole industry whereby an organisation can benchmark itself against the national performance of the industry and identify areas for improvement, that is, where they perform badly. However, these measures are specific to projects and offer very little indication as to the performance of the organisations themselves from a business point of view apart perhaps from the 'customer perspective'.

An outline of a typical performance management plan has been briefly outlined in Briefing Note 5.01.

Health, safety and welfare systems

In accordance with the construction phase health and safety plan with would have been prepared prior to commencement on site by the contractor at construction stage, it is the responsibility of the principal contractor to ensure adequate health and safety and welfares systems have been implemented. There are a number of initiatives for example Considerate Constructors Scheme (http://www.ccscheme.org.uk) which provides information guidance and monitoring and benchmark service to contractors with a view to improve H&S and welfare systems across the industry. Often in order to encourage clients to support these initiatives links are established with environmental performances (i.e. BREEAM credits).

Environmental statements

Environmental concerns will increasingly affect our projects. This is especially the case with the pressure to develop brownfield sites and reuse old sites. The cost of addressing contaminants or other environmental issues can add significant costs and increase the duration of project. Planning authorities are also more likely to instruct environment studies and restraints as part of the planning process, all of which must be incorporated into the project during the construction stage. It is the project manager that has overall responsibility to ensure compliance with these aims, objectives and constraints. The project manager will need to:

- Understand and act on the environmental impact assessment; see Briefing Note 2.03

- Ensure proper environmental advice is available.

- Ensure that the contractor is complying with the environmental statement; see Briefing Note 2.03

- Seek and ensure action by the contractor of any remedial actions should they be necessary to comply with environmental considerations.

Contractor's environmental management systems

The contractor must establish his own environmental management systems (EMS), but it is for the project manager to ensure that it is being managed properly and is progressing sufficiently to achieve all EMS objectives. Therefore the project manager should:

- Receive details of the contractor's EMS and the environmental plan (EP) specific to the project.

- Ensure that the contractor has set up all necessary procedures and structure to manage the EMS and implement the objectives of the EP.

- Check that the contractor's environment management plan matches the aims and objectives of the environmental statement.

- Agree with the contractor any further aims, specific targets or initiatives that will maximise sustainability of the project and minimise the detrimental impact of the construction process.

- Proactively monitor the progress of the contractor to maintain his proposals and objectives.

Compliance with site waste management plan regulations 2008

In April 2008, site waste management plans (SWMPs) became a legal requirement for all construction and demolition projects in England, for projects valued over £300,000. A SWMP provides a framework for managing the disposal of waste throughout the life of a construction project. In essence, it should contain the following information:

- ownership of the document

- information about who will be removing the waste

- the types of waste to be removed

- details of the site(s) where the waste is being taken

- a post-completion statement confirming that the SWMP was monitored and updated on a regular basis

- an explanation of any deviation from the plan

Generally, the SWMP will be instigated by the client at the pre-construction stage, where the designers will also have to provide the required information. At the construction stage the document becomes the responsibility of the principal contractor.

Monitoring of the works

Once the project is underway on the site, regular monitoring of progress is to be carried out by the project manager. There is a fine line as to how involved the project manager should become with the everyday issues facing the contractor, and thus the relationship, as mentioned previously, will determine the appropriate approach.

It is the project manager's responsibility to arrange from the outset progress meetings at regular intervals. During these meetings, the contractor will present a report as to progress on the site with any relevant design issues which will require resolving. If necessary, separate design meetings should also be set up. The reporting process to the project manager must not be restricted to the contractor but also to all designers and consultants. It is at these forums that the project manager must manage and ensure all parties are working together and achieving individual target dates for producing information and maintaining progress against the schedule.

Notwithstanding formal progress meetings, the project manager should also visit site regularly and spend limited time at the site discussing progress with site staff and chasing up the appropriate individuals for information and progress.

Reporting

A fundamental aspect of the project management role is the regular reporting of the current status of the project to the client. The project manager needs to ensure an adequate reporting structure and calendar is in place with the consultants and contractors. Frequency and dates of project meetings need to be coordinated with the reporting structure. Reporting is required for a number of reasons:

- to keep the client informed of the project status

- to confirm that the necessary management controls are being operated by the project team

- to provide a discipline and structure for the team

- as a communication mechanism for keeping the whole team up to date

- to provide an auditable trail of actions and decisions

Progress reporting should record the status of the project at a particular date against what the position should have been; it should cover all aspects of the project, identify problems and decisions taken or required, and predict the outcome of the project. The project manager needs to receive individual reports from the consultants and contractor and summarise them for the report to the client. The detailed reports should be appended as a record. Typical contents of a project manager's project report would contain the following:

- an executive summary

- legal agreements

- design status

- planning/Building Regulation status

- procurement status

- construction status

- statutory consents and approvals

- master development schedule and progress

- project financial report

- variation register update

- major decisions and approvals required

Trends shown visually are an excellent mode of conveying information to clients and senior management.

Public liaison and profile

The client would probably have set out their overall public relations and liaison strategy during the pre-construction stages of the project. A good local public relation, particularly during the construction stage, will help to improve the public's perception of the construction industry in general. Such activities or actions should include the following:

- ensuring that there is no local nuisance or negative impact arising from the project

- maintaining good housekeeping both on-site and in the immediate off-site area

- erecting informative scheme boards and public viewing platforms

- ensuring that the contractor takes part in a local or national 'considerate contractor' scheme

- taking awareness initiatives with local schools

- attending local public meetings to raise the profile of the project

- organising site visits for local schools, residents and business people

- partaking in local environmental schemes or issues

- being involved with fundraising for local charities or causes

Quality management systems

Quality in construction can be thought of as the satisfaction of a whole range of performance criteria owned by an interacting collection of stakeholders and mediated by a range of mechanisms running from regulation to market forces. Client satisfaction is the ultimate measure of construction quality, but that this will only be achieved if construction companies adopt a strong external orientation in order to address the full range of quality dimensions that impact on the client. Then there is a good prospect that the client will get satisfied users, satisfied statutory authorities, etc., as well as meeting their own direct ends.

In order to achieve and implement quality management systems, either an organisation-based approach or a project-based approach can be undertaken.

Typically, the quality management requirements will be developed at the strategy stage and will form part of the Project Execution Plan.

Commissioning and production of operation and maintenance manuals

Commissioning

The main commissioning and putting to work issues are covered in Stages 6 and 7. This section covers the construction stage.

The project manager should receive the contractor's commissioning schedule within the early stages of construction in order to satisfy himself that it is properly coordinated with the building works schedule. (As an example, the balancing of the heating and air conditioning system can only take place when the building envelope and internal spaces have been secured.) A problem may occur in that in many instances the building services contractor is a subcontractor to the main contractor and that the subcontract may not be in place at this early stage of construction. In this case, the main contractor will need to identify the logic and sequence of the commissioning.

Operation and maintenance manuals

The CDM Regulations now cover the operation and maintenance manuals, and it is the CDM coordinator's role to ensure that they are delivered as part of the health and safety file. These manuals should also include details of the complete building with input from all of the design team. The project manager has to monitor the progress that the CDM coordinator is making on assembling these files and, if needs be, ensure that all necessary actions are taken to expedite their completion with active cooperation from the contractor. It is a legal responsibility of the contractor to cooperate with the CDM coordinator and to comply with any reasonable request from the CDM coordinator in order to enable completion of the health and safety file including the operating and maintenance manuals.

BIM strategy

Once operations have commenced on site, the project BIM should ideally be at full detailed coordinated construction status. This will include any input in terms of design geometry and data from supply chain, that is, subcontractors with the design responsibility and product suppliers.

This is a major change from traditional 2D construction information environments, where design development work and construction operations can have substantial overlap in programming terms.

In using BIM to its full potential, the construction BIM geometry and model should be complete or virtually so before operations start on site. This means changing the emphasis on design input to the contract mobilisation period and pre-construction activities to enable effective supply chain involvement.

In this way, construction operations can commence with the delivery team having in their possession most if not all of the information they need to deliver the project successfully. This information will have been clash checked, coordinated and validated for compliance with statutory regulations and the client's requirements, giving more certainty and predictability to the project outcomes.

During the course of construction, the BIM is updated with changes resulting from the site operations, and any other inputs such as client changes and design changes.

The site team will use 'field BIM', with site managers accessing the BIM via handheld tablets and while walking round the project they can update the BIM with progress, snagging and any other information available. In this way, the BIM is kept up to date, so that at completion of the construction stage, the BIM is an accurate and complete representation of the constructed project.

Briefing Note 5.01   Performance management plan

Performance management should be an integrated part of a development project from its definition through to monitoring and review. Where it is not possible to establish direct 'cause and affect' linkages or precise measures of performance, interim measures such as key performance indicators (KPIs) are often used. These could be trends over time, value to the customer, awareness of product or service.

Objectives

The purpose of a performance management plan (PMP) is to set out the principles and targets for a schedule against which it delivers its outputs, outcomes and benefits. The plan also defines how the performance criteria will be measured and plans for any divergence management. The plan contains details of the performance management process, performance measurements and the performance information required to establish and monitor delivery.

Performance management process

The performance management process outlines the activities to set direction, which uses performance information to manage better, demonstrates what has been accomplished and sets actions to improve. Performance metrics may be defined using the SMART test (Specific, Measurable, Attainable, Relevant and Timely).

Performance measurements should indicate milestones for measuring progress against goals (may be at the key decision stages), against target levels of intended accomplishment (target objectives) and against third parties. Measures may need to change as progress is made. Measurement criteria may be defined using the FABRIC test (Focused, Appropriate, Balanced, Robust, Integrated and Cost-effective).

Performance information includes the data, their characteristics, quality, sources and contribution to a measure.

Checklist for PMP

Quality criteria for a PMP include:

- Are the objectives, outputs, outcomes or benefits against which to set and monitor performance or achievement of targets clearly defined?

- Can the performance measures be assessed against key objectives?

- Are the performance measures clearly defined, together with target values?

- Is the approach for managing performance is complete and does it contain all key elements as a cycle of activities?

- Do the measures and metrics criteria meet pre-defined tests of SMART and FABRIC (if applicable)?

- Is the periodicity of measurement clearly defined?

- Have all standards or techniques to be used for measurement are defined?

- Are the sources of performance information of adequate quality?

- Is the proposed performance information reliable and/or independently validated?

- Is the approach to investigation and corrective action to improve unsatisfactory performance clearly defined?

- Is there an outline of the management organisation and process?

- Are the resources to collect and analyse performance information clearly defined?

- Are the roles and responsibilities clearly defined?

Suggested contents for the PMP

The key elements of performance management plan will describe a cycle of activities and their outputs:

- Strategy: defining the aims and objectives of the organisation.

- Selection of performance measures: identifying the measures which support the quantification of activities over time.

- Selection of targets: quantifying the objectives set by management, to be attained at a future date.

- Delivery of performance information: providing a good picture of whether an organisation is achieving its objectives.

- Reporting information: providing the basis for internal management monitoring and decision making, and the means by which external accountability is achieved.

- Action to improve: taking action to put things right; feeding back achievements into the overall strategy of the organisation.

6 Testing and commissioning

Stage checklist

Key processes:	Commissioning services
	Commissioning documentation
	Quality management and control
Key objective:	'Is it performing as designed?'
Key deliverables:	Commissioning documentation
Key resources:	Client team
	Project manager
	Design team
	CDM coordinator
	Constructor team
	Commissioning team

Stage processes and outcomes

Part of the construction stage, this stage confirms that the building services systems have been installed in compliance with the design, have been fully tested and have been proven to be fully functional.

Outcomes:

- Commissioning strategy and schedule

- Commissioning outcomes in accordance with contractual requirements

- O&M manuals

- Certifications and warranties

- Records of testing and witnessing

- Training of client staff

- Update Health and Safety Plan and File

- Performance certificates

It must be stressed that the sequencing of this stage does not mean that the activities involved only take place at the end of the construction stage. Commissioning is a very important part of the construction process and must be addressed and considered

Code of Practice for Project Management for Construction and Development, Fifth Edition. Chartered Institute of Building.
© Chartered Institute of Building 2014. Published 2014 by John Wiley & Sons, Ltd.

very early on within the project. The following are suggested activities that must be considered well before this stage:

- Decide the most appropriate time within the project to appoint the commissioning contractor and their role/scope of work.

- Where appropriate, appoint a commissioning contractor to review the design drawings and working drawings to ensure commissionability.

- Ensure consultants clearly identify testing and commissioning requirements.

- Ensure consultants/client identify performance/environmental testing requirements.

- Ensure that the master development schedule includes sufficient time to undertake the specified commissioning and, in particular, the additional time required for any performance/environmental testing and statutory testing to authorities.

- Clearly identify the method of presenting, recording and electronically storing 'as-installed' information.

- Although not strictly part of commissioning, ensure that the requirement for specialist maintenance contracts for equipment is carefully considered prior to awarding tenders for such equipment.

Project manager's duties and responsibilities

The project manager's objective is to ensure that the commissioning of the separate systems is properly planned and executed, so that the installation as a whole is fully operational at handover without delay to the programme and that any fine-tuning necessary after handover is carried out in liaison with the client and/or user.

Commissioning generally

Commissioning is carried out in four or sometimes five distinct parts:

- static testing of services
- dynamic testing of services
- performance testing of services (not always undertaken)
- undertaking statutory tests for various authorities
- client commissioning

Note that performance testing also includes environmental testing. The first four items, services testing, commissioning, performance testing and statutory tests, are part of the construction design and installation phases of the project. Client commissioning is an activity predominantly carried out by the client's personnel assisted, where required, by the consultants. This is dealt with in Stage 7. The services testing and commissioning process objectives and main tasks are as described within this chapter.

Procurement of commissioning services

Smaller projects

There are many ways to procure the commissioning specialist. On smaller projects, via the main contractor, the mechanical and electrical subcontractors are most likely to be responsible for the testing and commissioning of their installations. Electrical contractors will normally use their own resources, except where specialist items of equipment require the manufacturer to assist with their testing. Mechanical contractors will usually appoint a commissioning specialist to work on their behalf. Again, where specialist items of equipment are installed, the mechanical contractor will request the manufacturer to assist with testing where appropriate. However, it should be noted that often these commissioning specialists are no more than balancing engineers. Where more complicated systems are involved or specific commissioning and performance tests are required, specialist inputs will be required. Careful specification of the requirements within the design documentation is required when tendering the installation work. This is all too often ignored or given insufficient time and effort which inevitably creates problems later in the construction process.

Larger projects

On larger projects, the method of procuring the commissioning specialist can take many forms. In traditional forms of contract, it can again be via the main contractor/services contractor; however, in construction management or similar forms of contract, a specialist commissioning contractor is often appointed. This commissioning contractor normally fulfils one of two roles: the role of managing the testing and commissioning process (the actual work being done by the installation contractors as detailed for small projects above), or the role of undertaking the commissioning work. In this latter role, the point of delineation for testing/commissioning between the installation contractor and the commissioning contractor is usually at the end of static testing and the start of dynamic testing. See the following for a definition of these terms. This latter role is gaining in popularity for the following reasons:

- It provides a degree of independence to the commissioning process.

- The commissioning contractor is likely to be under the control of the client/construction manager/main contractor and reports directly to them, giving greater control and transparency to the process.

In either role, the benefit to the project is that the commissioning contractor can be brought into the project very early to manage the whole testing and commissioning process.

Role of the commissioning contractor

What follows are some of the activities that can be included within the scope of work for the commissioning contractor:

- Review the design drawings near the end of design to ensure familiarity with the design intent and to add their expertise in to the commissionability of a scheme.

- Ensure that the testing and commissioning is correctly specified in the tender documentation.

- Review the services contractor's working drawings for commissionability.

- Set up the testing and commissioning documentation to create consistency between the various contractors.

- Define the method, media type, style and content of the as-installed information to create consistency between the various contractors.

- Manage the specialist equipment manufacturers' tests.

- Liaise with building control and other organisations to witness relevant statutory tests (including insurer's tests).

All of these functions are often given insufficient thought on projects, so if they are not to form part of the commissioning contractor's brief, then it should be recognised that some other part of the project team should undertake this work.

The testing and commissioning process and its programming

Flowcharts relating to the various stages of testing, commissioning and performance testing are given in Figure 6.1 and Figure 6.2. It is important for the project manager to understand the differences between the terms testing, commissioning and performance testing, and for him to ensure that the programme has sufficient time within it to enable these activities to be undertaken. Unfortunately, with this stage of the project being so close to handover, there is often pressure to gain time by shortening the testing, commissioning and performance/environmental testing schedule. This should be strongly resisted. Rarely, if ever, after the project will such an opportunity

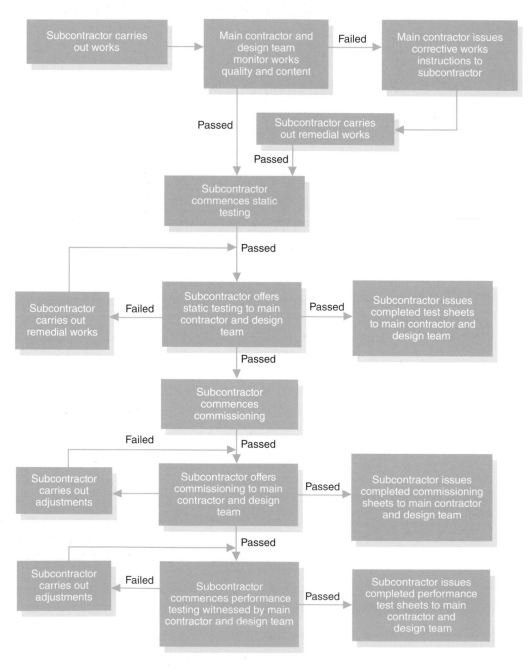

Figure 6.1 Small project installation testing and commissioning process and sign off.

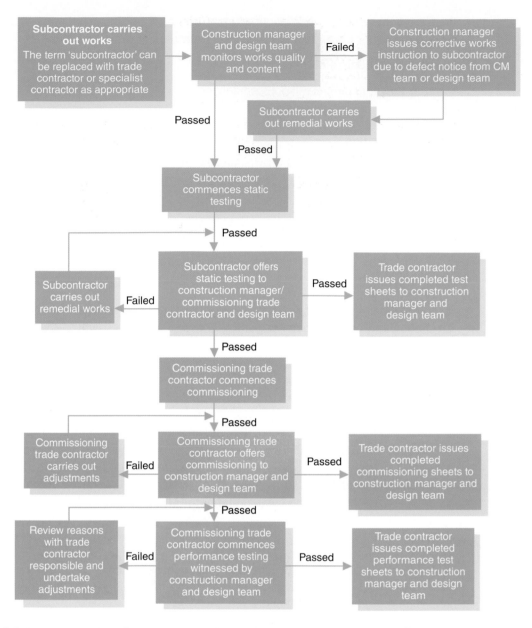

Figure 6.2 Large project installation testing and commissioning process and sign off.

exist to fully test the services to ensure that they work individually, as a system, and, that they work under part-load and full-load conditions. Many issues with respect to the underperformance of services within an occupied building can be related back to either insufficient quality in the testing and commissioning, or insufficient time to test and commission.

It should also be borne in mind that various statutory services will need to be demonstrated to building control (or the relevant government department if a Crown building) and insurers. Time should be allowed for within the programme since these activities are often taken as separate tests after the main commissioning has been undertaken.

Differences between testing and commissioning

Testing

During the services installation, various testing will be undertaken known as 'static testing'. This testing is normally undertaken to prove the quality and workmanship of the installation. Such work is undertaken before a certificate is issued to 'liven' services whether electrically or otherwise. Examples of systems tested are:

* pressure testing ductwork and pipework

* undertaking resistance checks on cabling

Commissioning

On completion of the static testing, dynamic testing commences: this being the commissioning. Commissioning is undertaken to prove that the systems operate and perform to the intended design and specification. This work is extensive and normally commences by issuing a certificate permitting the installation to be made 'live', that is, electrical power on. After initial tests of phase rotation on the electrical installation and checking fan/pump rotation (in the correct direction), the more recognised commissioning activities of balancing, volume testing, load bank testing and soon begin.

Performance testing

On completion of the commissioning, performance testing can commence. Some may not distinguish between commissioning and performance testing. However, for the purposes of the master development schedule, it is worth distinguishing the difference between commissioning plant as individual systems and undertaking tests of all plant systems together, known as performance testing (and including environmental testing). Sometimes this performance testing is undertaken once the client has occupied the facility, for example, for the first year because systems depend on different weather conditions. In such cases, arrangements for contractor access after handover to fine-tune the services in response to changing demands must be made. However, for some facilities it is desirable, if not essential, to simulate the various conditions expected to prove that the plant systems and controls operate prior to handover, for example, in the case of computer server rooms.

Main tasks to be undertaken

To assist the project manager, the following has been provided to summarise the main tasks to be carried out during the three main stages of pre-construction, construction and post-construction.

Pre-construction

The following items will need to be confirmed:

* The consultants/client recognises engineering services commissioning as a distinct phase in the construction process which has an important interface with client commissioning (see Stage 7).

* The relevant consultants identify all services to be commissioned and define the responsibility split for commissioning between designers, contractor, manufacturer and client. Responsibility for specialised plant/services is defined early, particularly 'wear and tear' and the cost of consumables, fuel, power, water, etc.

* The services designers, and commissioning contractor if relevant, audit the final layout drawings to ensure that they make provision for the systems to be commissioned in accordance with the relevant codes of practice.

* The consultants/client, and commissioning contractor if relevant, identify all required statutory and insurance approvals relating to services commissioning, and see that plans are made for meeting requirements and obtaining the approvals (see Stage 7).

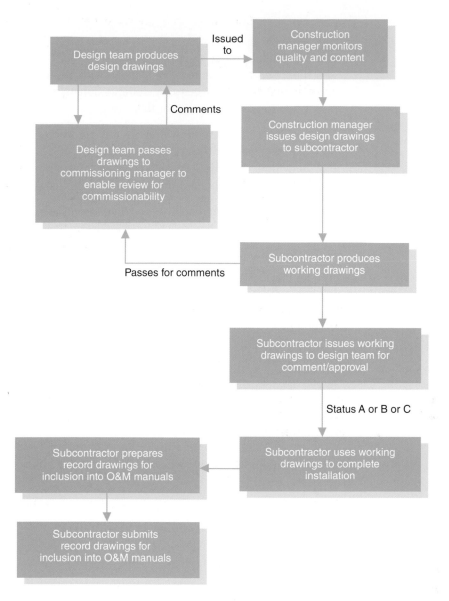

Figure 6.3 Project drawing issue flowchart.

- That the client understands the importance of the presence of the client's own maintenance/engineering department/maintenance contractor during the commissioning process

- That the client considers whether an aftercare engineer needs to be appointed to support the client/user in the first 6–12 months of occupancy.

- There is a schedule showing the timescale and sequence of commissioning and testing and handover events, system by system. This is essential.

- There should be a person nominated by the client to act on behalf of the client in terms of commissioning activities, who should be a member of the client's team as defined in Stage 7 – Completion, Handover and Operation. This does not preclude more than one person having the benefit of witnessing the commissioning process.

- The contract documents *must* make adequate provision for testing, commissioning and performance testing. Confirmation on warranties, defects period, environmental testing at this stage will set out what level of commissioning is required and ensure that responsibility for plant and systems is still with the contractor during an extended commissioning period (see Stage 7) (Figure 6.3).

Construction and post-construction

- The consultants must inspect the work for which they have design responsibility, and report on progress and compliance with contract provisions, highlighting any corrective action necessary. A commissioning management specialist may be appointed to carry out much of this work.

- There must be confirmation that all the contractor's working schedules include commissioning activities and that they are properly related to preceding construction activities. Activities must be complete, timings reasonable and compatible with planned handover, and properly related to preceding activities.

- Coordination of the consultants' arrangements is required for client involvement in, or observation of, contractor's commissioning against contract arrangements.

- Monitoring and reporting progress of commissioning will be carried out to ensure that activities start as scheduled and that the requirements for completion before handover are met. Corrective action will have to be initiated as necessary. It is important that commissioning activity durations do not become eroded due to late or incomplete construction work.

- All 'completed construction' documents should be in place before commissioning an individual system commences, for example, cleaning out, testing the electrical power and controls to it. Also, the requirements of 'permits to work' and health and safety should be met; and responsibility for insurance should be clearly defined.

- Statutory/insurance tests should be arranged and undertaken, witnessed by the relevant authority, for example, building control, utility companies, fire brigade, insurers, etc.

- Commissioning records, for example, test results, calibration requirements, certificates and checklists must be properly maintained and copies bound into the operating and maintenance manuals or in separate commissioning manuals to form part of the official handover documentation.

- Operating and maintenance manuals, 'as-installed' record drawings and the client's staff training have to be provided by the contractor as required under the contract, although it is recommended that these are fully coordinated by others, for example, the commissioning contractor, if appointed.

- Adopting agreed structure and software for operating and maintenance manuals with copy disks provided for ease of updating.

- Record drawings being provided in a computer-aided drawing format for ease of updating.

- Using video recordings during client-training sessions for subsequent repeat visual reference and to assist new maintenance staff advance along the learning curve (Figure 6.4 and Figure 6.5).

Seasonal commissioning

Seasonal commissioning involves re-commissioning heating systems in winter and mechanical cooling systems in summer. But seasonal commissioning may also be applied to other systems, such as motorised windows and active solar-shading devices – or indeed any other any building system affected by seasonal changes. Ideally, the original project team (or independent commissioning engineer, if appointed) should remain engaged to perform the seasonal commissioning.

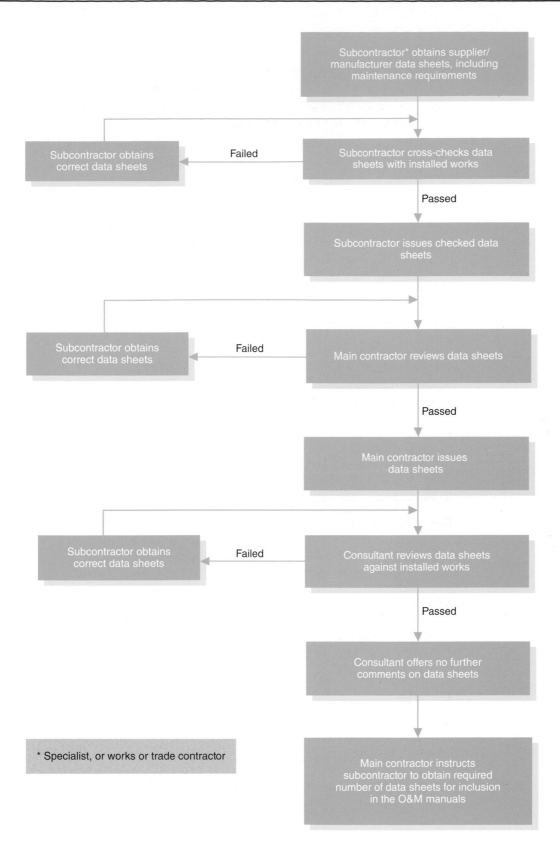

Figure 6.4 Services installation, testing and commissioning data sheets flowchart.

Many environmental assessment schemes award credits for the adoption of seasonal commissioning. However, in practice, sometimes it may difficult to distinguish seasonal commissioning activities from continuous commissioning, or activities that were following on after handover due to programme, logistics or contractual issues.

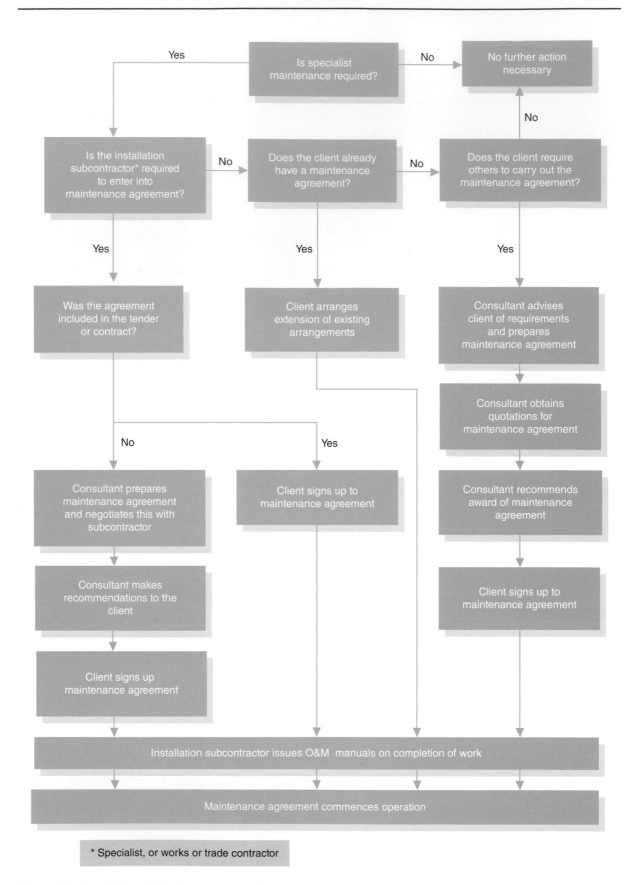

Figure 6.5 Specialist maintenance contracts flowchart.

Commissioning documentation

O&M manual (building owner's manual)

The building owner's manual, or operation and maintenance manual (O&M manual), contains the information required for the operation, maintenance, decommissioning and demolition of a building.

The building owner's manual is prepared by the contractor with additional information from the designers, suppliers and the CDM coordinator. It is a requirement that is generally defined in the preliminaries section of the tender documentation where its contents will typically be outlined, although there may be additional requirements regarding mechanical, electrical and any other associated services in the mechanical and electrical specification.

A draft version of the document should be provided for the client/project manager as part of the handover procedure prior to certifying practical completion. The final document is not usually available in full form until after practical completion, as commissioning information often needs to include summer and winter readings taken in the fully occupied building. The preliminaries may require several copies of the building owner's manual and might often require an electronic version.

The building owner's manual might include:

- A description of the main design principles.

- Details of the building's construction (such as finishes, cladding, doors and windows, roof construction, and so on).

- As-built drawings and specifications.

- Instructions for its operation and maintenance (including health and safety information and manufacturers' instructions for efficient and proper operation).

- An asset registers of plant and equipment.

- Commissioning and testing results.

- Guarantees, warranties and certificates.

- Particular requirements for demolition, decommissioning and disposal.

Much of this information will already exist in one form or another, so preparing the building owner's manual might simply be a matter of compiling and assembling its components.

Over the life of the buildings, the building owner's manual may be developed to reflect changes that take place to the fabric of the building or its systems, along with details of maintenance that has taken place.

The building owner's manual is different from the health and safety file which only needs to include information important to health and safety and does not need to include information about the construction process, contractual information or information about the normal operation of the completed structure (see Managing health and safety in construction: Construction (Design and Management) Regulations 2007 Approved Code of Practice page 60).

The building owner's manual may also include a non-technical 'building user's guide' with information for users about environmental controls, access, security and safety systems, etc., and record any training provided to the building users and the facilities management team for operating and maintaining these functionalities.

As-built documentation

'As-built documentation' is a fundamental part of the Works Contract, without these documents the Contract will not be deemed as complete. An advance copy of the 'As-built documentation' is submitted to the project manager prior to the proposed completion of the works. The project manager will appraise the submission (typically through the designers) and return the copy to the contractor together with any comments. The contractor shall ensure that the comments submitted by the project manager are incorporated into the final 'As-built documentation'.

The as-built documentation, which will include the as built drawings and material data sheets, is often incorporated within the building owner's manual.

If there have been specification changes during the project, then the as-built specifications including material data sheets are also recorded within this section. As-built Information will include:

- civil engineering works including below ground drainage, substructure and foundation works

- structural information

- architectural information

- building services information

- specialist design and installation information

Health and safety file

The Construction (Design and Management) Regulations (CDM regulations) require that the CDM coordinator '…*prepare, where none exists, and otherwise review and update a record ("the health and safety file") containing information relating to the project which is likely to be needed during any subsequent construction work to ensure the health and safety of any person, including the information provided in pursuance of regulations 17(1), 18(2) and 22(1)(j)*', and that at the end of the construction phase, they pass the health and safety file to the client.

The health and safety file need only include information important to enable future construction work including cleaning, maintenance, refurbishment, alterations and eventual demolition safety. If it contains all information about the building then genuinely important safety issues may simply be overlooked. It does not need to include information about the construction process (which may be included in the construction phase plan), unless it may affect future works. It does not need to include contractual information or information about the normal operation of the completed structure (which may be included in the building owner's manual).

The health and safety file may contain:

- a description of the project

- a description of any residual hazards that should be managed

- the structural principles of the design

- identification of any hazardous materials used

- procedures for the removal or dismantling of installed plant and equipment

- information about cleaning and maintenance equipment

- a description of significant services and their location

- information and as-built drawings of the structure, plant and equipment

The health and safety file must be kept up to date and is normally kept for the lifetime of the building, meaning that it should be passed on to the new owners if the building is sold, and the new owners should be informed of its purpose and importance.

There are no restrictions to the format that it has to be kept in, but it would be wise to ensure it is backed up. To be useful, the health and safety file has to be kept up to date.

If the building or part of it is leased, then the health and safety file to must be made available to the leaseholder. If there are multiple leaseholders, then those parts of the health and safety file relevant to the part of the building leased by each lease-holder must be made available to them. In multi-occupancy situations, for example, where a housing association owns a block of flats, the owner should keep and main-tain the file, but ensure that individual flat occupiers are supplied with health and safety information concerning their home.

Occupier's handbook

The purpose of this handbook is to:

- outline the services that the building occupiers can expect relative to health and safety and welfare

- highlight those areas where occupier cooperation is required by the property owner and/or their facilities managers relative to health and safety and welfare issues

- outline the minimum health and safety and welfare standards that are expected of the occupiers to observe

- provide guidance to occupiers on measures that can be taken to safeguard prop-erty, by the proper control of contractors

- provide general information regarding the Insurance's arranged by the property owner and the procedures for reporting any loss, damage, accident or other incidents

A suggested table of contents for a typical occupier's hand book is outlined in Briefing Note 6.03.

BIM strategy

Building services can represent more than 30% of a project's capital cost. Therefore, the design, integration and coordination of the services package with the structural and architectural elements of the design is a critical success factor.

The BIM platform will have been used to integrate the services design into the project, with attached data sets including performance and technical data, and any maintenance and operations information.

Using the Soft Landings process or similar, planning for handover and commissioning should begin at the latest mid way through the construction period.

Commissioning input should be obtained into the servises design activities in order to ensure that the systems are as simple to commission and test as possible.

The project BIM can be used to walk through the commissioning process and help the team to understand the requirements and sequence. Also, BIM can be used to iso-late and visualise particular systems. It is also possible to simulate the building's designed performance and determine the design outputs for carbon emissions, energy consumption, use of resources, etc.

Briefing Note 6.01   Contents of the health and safety file

The health and safety file should contain the information needed to allow future construction work, including cleaning, maintenance, alterations, refurbishment and demolition to be carried out safely. Information in the file should alert those carrying out such work to risks, and should help them to decide how to work safely. The file should be useful to:

(a) clients, who have a duty to provide information about their premises to those who carry out work there

(b) designers during the development of further designs or alterations

(c) CDM coordinators preparing for construction work

(d) principal contractors and contractors preparing to carry out or manage such work

The file should form a key part of the information that the client, or the client's successor, is required to provide for future construction projects under regulation 10. The file should therefore be kept up to date after any relevant work or surveys.

The scope, structure and format for the file should be agreed between the client and CDM coordinator at the start of a project. There can be a separate file for each structure, one for an entire project or site, or one for a group of related structures. The file may be combined with the Building Regulations Log Book, or a maintenance manual providing that this does not result in the health and safety information being lost or buried. What matters is that people can find the information they need easily and that any differences between similar structures are clearly shown.

Obligations

Clients, designers, principal contractors, other contractors and CDM coordinators all have legal duties in respect of the health and safety file:

(a) CDM coordinators must prepare, review, amend or add to the file as the project progresses, and give it to the client at the end of project

(b) clients, designers, principal contractors and other contractors must supply the information necessary for compiling or updating the file

(c) clients must keep the file to assist with future construction work

(d) everyone providing information should make sure that it is accurate, and provided promptly

A file must be produced or updated (if one already exists) as part of all notifiable projects. For some projects, for example, redecoration using non-toxic materials, there may be nothing of substance to record. Only information likely to be significant for health and safety in future work need be included. It is not necessary to produce a file on the whole structure if a project only involves a small amount of construction work on part of the structure.

The client should make sure that the CDM coordinator compiles the file. In some cases, for example, design and build contracts, it is more practical for the principal contractor to obtain the information needed for the file from the specialist contractors. In these circumstances, the principal contractor can assemble the information and give it to the CDM coordinator as the work is completed.

It can be difficult to obtain information for the file after designers or contractors have completed their work. What is needed should be agreed in advance to ensure that the information is prepared and handed over in the required form and at the right time.

The contents of the health and safety file

When putting together the health and safety file, it is necessary to consider including information about each of the following where they are relevant to the health and safety of any future construction work. The level of detail should allow the likely risks to be identified and addressed by those carrying out the work:

(a)   a brief description of the work carried out

(b)   any residual hazards which remain and how they have been dealt with (e.g. surveys or other information concerning asbestos; contaminated land; water bearing strata; buried services, etc.)

(c)   key structural principles (e.g. bracing, sources of substantial stored energy – including pre- or post-tensioned members) and safe working loads for floors and roofs, particularly where these may preclude placing scaffolding or heavy machinery there

(d)   hazardous materials used (e.g. lead paint; pesticides; special coatings which should not be burnt off, etc.)

(e)   information regarding the removal or dismantling of installed plant and equipment (e.g. any special arrangements for lifting, order or other special instructions for dismantling, etc.)

(f)   health and safety information about equipment provided for cleaning or maintaining the structure

(g)   the nature, location and markings of significant services, including underground cables; gas supply equipment; fire-fighting services. etc.

(h)   information and as-built drawings of the structure, its plant and equipment (e.g. the means of safe access to and from service voids, fire doors and compartmentalisation, etc.)

The file does not need to include things that will be of no help when planning future construction work, for example:

(a)   the pre-construction information, or construction phase plan

(b)   construction phase risk assessments, written systems of work and COSHH assessments

(c)   details about the normal operation of the completed structure

6.01 Briefing Note

(d)   construction phase accident statistics

(e)   details of all the contractors and designers involved in the project (though it may be useful to include details of the principal contractor and CDM coordinator)

(f)   contractual documents

(g)   information about structures, or parts of structures, that have been demolished – unless there are any implications for remaining or future structures, for example, voids

(h)   information contained in other documents, but relevant cross-references should be included

Some of these items may be useful to the client, or may be needed for purposes other than complying with the CDM Regulations, but the Regulations themselves do not require them to be included in the file. Including too much material may hide crucial information about risks.

Storing the file after the work is complete

To be useful the file needs to be kept up to date, and retained for as long as it is relevant – normally the lifetime of the structure. It may be kept electronically (with suitable backup arrangements), on paper, on film or any other durable form.

Where clients dispose of their entire interest in a structure, they should pass the file to the new owners and ensure that they are aware of the nature and purpose of the file. Where they sell part of a structure, any relevant information in the file should be passed or copied to the new owner.

If the client leases out all or part of the structure, arrangements need to be made for the health and safety file to be made available to leaseholders. In some cases, the client might transfer the file to the leaseholder during the lease period. In other cases, it may be better for the client to keep the file, but tell leaseholders that it is available. If the leaseholder acts as a client for future construction projects, the leaseholder and the original client will need to make arrangements for the file to be made available to the new CDM coordinator.

In multi-occupancy situations, for example, where a housing association owns a block of flats, the owner should keep and maintain the file, but ensure that individual flat occupiers are supplied with health and safety information concerning their home.

A development may include roads and sewers that will be adopted by the local authority or water company. It is generally best to prepare separate files covering each client's interests.

Further information is available in the Health and Safety Executive website and the Approved Code of Practice (ACOP).

Briefing Note 6.02   Contents of building owner's manual

Section 1   Detail all parties to the project – Project Directory

1.1   The Client – including the person responsible for the project:
    Address
    Telephone numbers
    E-mail address
    Contract reference
    Project start and finish dates

1.2   The Contractor – including the senior member of staff responsible for the project:
    Address
    Telephone numbers
    E-mail address
    Contractors project/job number
    Emergency telephone number, 24 hour, for the duration of the defects liability period

1.3   The Sub-Contractors – including the senior member of staff responsible for the project,
    Address
    Telephone numbers
    E-mail address
    Contractors project/job number
    Emergency telephone number, 24 hour, for the duration of the defects liability period

Section 2   Brief description of the project

This section should outline the scope and the outline description of the project.

Section 3   Maintenance instructions and schedules

This section shall contain maintenance schedules indicating
  (a) recommended periods between testing, servicing and inspection for all systems and equipment based upon a five-year period
  (b) spares requirements for the installation and equipment based upon a five-year period
  (c) reference shall be included to relevant British Standards, Codes of Practice and Manufacturers requirements

Included in this section shall be a brief description of operation and maintenance procedures for
- fire alarm systems
- ventilation systems, including controls
- comfort cooling, including controls
- process chilled water systems and controls
- compressed air systems
- security systems, including card access
- lighting systems, including controls
- heating systems, including controls
- power systems, including controls
- any miscellaneous systems and equipment
- soil and drainage

Section 4   Equipment details

The section shall start with a schedule of all M&E equipment used on the project subdivided into

- switchgear
- sub-distribution boards
- power systems
- luminaries
- electrical accessories
- data equipment and accessories
- cables/wiring
- fire alarm equipment
- security equipment
- miscellaneous electrical
- equipment
- boilers
- supply and extract AHU
- chillers
- noise attenuation
- valve/strainers, etc.
- filters
- lagging
- pumps
- filters
- pipes

The remainder of this section shall contain manufactures data sheets/catalogue extracts for the items as contained and in the order of the schedule. Where a manufacturer sheet contains more than one component, the contractor shall ensure that the individual item(s) are clearly indicated on the sheet. Manufacturer's warranty details, that is, duration and any conditions applicable.

Section 5   Project drawings (as-built drawings)

The hard copies of the drawings shall be submitted within individual plastic pocket-style inserted within the ring binders. Drawings scale (1:1) of the installation and arrangements on re-writable CD-ROM, created using AutoCAD Lt20XX, or AutoCAD release 20XX, later versions of AutoCAD may be used with prior agreement with the engineer. The disc(s) shall be installed within a purpose made ring binder disc pocket.

Drawings shall be saved as unlocked .dwg (suitable for incorporation into the universities system for future editing). Note that 'xref' drawings shall not be submitted. Ensure that all drawings are 'purged' before compression, zipping or sending.

Section 6   Test and commissioning information

The hard copies of the Test and Completion Certificates plus Commissioning Schedules shall be submitted within individual plastic pocket-style inserts within the ring binders.

- Original copies of all test certificates
- Electrical Test Certificates
- Electrical Completion Certificates
- Fire Alarm Test Certificates, including audibility schedule
- Fire Alarm Cabling Test Result Schedule
- Fire Alarm Completion Certificate
- Emergency Lighting Test Certificates
- Emergency Lighting Completion Certificate
- Data/Telecommunication Test Schedules
- Data/Telecommunication Completion Certificate
- Commissioning + Test Schedules and Certificates for Miscellaneous Systems and Equipment
- Vibration data/static deflections

(Continued)

- Air flow rates and pressure drops
- Water flow rates and pressure drops
- Static pressure tests
- Building leakage test results
- Noise-level data

Section 7   Guarantees and warranties

Lists of all guarantees and warranties applicable to the project.

Section 8 Health and safety file

This should contain the information as required under the CDM Regulations, including information necessary for demolition, decommissioning and disposal.

Briefing Note 6.03 Contents of occupier's handbook

INTRODUCTION

1. Health and safety statement

2. Property owner/manager's services

3. Cooperation on health and safety matters

   3.1   General

   3.2   Evacuation procedures

   3.3   Control of substances hazardous to health

   3.4   Housekeeping

   3.5   Visitors

   3.6   Waste re-cycling and disposal

4. Occupiers obligations

   4.1   General

   4.2   Asbestos

   4.3   Bomb threats and suspicious objects

   4.4   Building alterations

   4.5   Contamination

   4.6   Disability access

   4.7   Electrical safety

   4.8   Environmental

   4.9   Fire safety

   4.10 First aid

   4.11 Floor loading

   4.12 Gas safety

   4.13 Highly flammable substances

   4.14 Pest control

   4.15 Security

   4.16 Traffic routes

   4.17 Unoccupied premises

   4.18 Water leakage

   4.19 Waste disposal (by occupiers)

   4.20 Windows

5. Control of contractors

   5.1   General

   5.2   Contract conditions

   5.3   Management of contractors

(Continued)

6.03 Briefing Note

7 Completion, handover and operation

Stage checklist

Key processes:	Planning and scheduling handover
	Handover procedures
	Operational commissioning
	Client occupation
Key objective:	'How do we use the building?'
Key deliverables:	Handover documentation
	Health and safety file
Key resources:	Client team
	Project manager
	Design team
	CDM coordinator
	Constructor team
	Commissioning team
	Occupation and maintenance team

Stage process and outcomes

This stage consists of the formal transfer of the completed facilities from the project team to the client, this may be a single event or phased over a period of time. It involves ensuring the client has the knowledge and capability to operate the new facilities. This stage also includes the occupation and use of the facilities by the client organisation.

Outcomes:

- handover strategy and procedures

- final inspections, physical handover and demobilisation

- Practical Completion Certificate

- schedule of any outstanding works or exclusions

- record of facilities as built

- end user/client familiarisation and training

- transfer of building insurances from contractor to client (if applicable)

- completion of contractors final account

- dispute resolution (if necessary)

- defects liability management

Code of Practice for Project Management for Construction and Development, Fifth Edition. Chartered Institute of Building.
© Chartered Institute of Building 2014. Published 2014 by John Wiley & Sons, Ltd.

Planning and scheduling handover

The overall objective is to schedule the required activities to achieve a coordinated and satisfactory completion of all work phases within the cost plan. This has to be meshed with the logistical planning of the client's occupation coordinator and any accommodation schedule of work to be completed prior to occupation.

Generally, construction projects can be subject to phased (sectional), partial as well as practical completion. The relevant procedures applied depend on the nature and complexity of the project, and/or requirements of the users. In effect, phased completion means the practical completion for each specific phase of construction. However, this must not:

- prevent or hinder any party from commencing, continuing or completing their contractual obligations

- interfere with the effective operation of any plant or services installations

In cases of phased completion handover, the client/end-user is usually responsible for insuring the works concerned. On practical completion handover, the whole of the insurance premium becomes the client/end-user responsibility.

Procedures

The actual practical completion and handover procedures applicable to a specific project will be detailed by the project manager in the Project Execution Plan (see Stage 3) for the project concerned (see Briefing Notes 7.05 and 7.06 for typical examples). However, the main aspects of completion and handover will generally cover the minutiae of the following activities:

- Preparation of lists identifying deficiencies, for example, unfinished work, weather damage and materials, goods and workmanship not in accordance with standards.

- All remedial and completion work carried out within the specified time under the direct supervision of nominated, qualified and experienced personnel.

- Monitoring and supervising completion and handover against the schedule.

- The provision of the required number of:

  - copies of the CDM health and safety file

  - 'as-built' and 'installed' record drawings, plans, schedules, specifications, performance data and tests results

  - commissioning and test reports, calibration records, operating and maintenance manuals, including related health, safety and emergency procedures

  - planned maintenance schedules and specialist manufacturers' working instructions

- Monitoring proposals for the training of engineering and other services staff and assistance in the actual implementation of agreed schemes.

- Ensuring that handover takes place when all statutory inspections and approvals are satisfactorily completed but does not take place if the client/end-user cannot have beneficial use of the facility, that is, not before specified defects are made good, indicating likely consequences and drawbacks of premature occupation.

- Setting up procedures to monitor and supervise any post-handover works, which do not form part of the main contract, and to monitor the defects liability period.

- Initiating, in close cooperation with the relevant consultants, cost-off setting measures in cases of difficulties with completing outstanding works or making good any defects.

- Monitoring progress of final accounts by assisting in any controversial aspects or disputes, and by ascertaining that draft final accounts are available on time and are accurate.

- Reviewing progress at regular intervals, to facilitate a successful final inspection, and the issuing of a final certificate.

- Establishing the plan for post-completion project evaluation and feedback from the parties to the contract for the post-completion review project close-out report.

Client commissioning and occupation

Having accepted the constructed structure from the contractor at practical completion, the client has to finally prepare the facilities ready for occupation. This stage of the project lifecycle comprises three major groups of tasks: client accommodation works, operational commissioning and migration.

- In order to allow as much time as possible for the client organisation to develop their detailed requirements, or to reflect their latest business 'shape', it is common for the client to organise a further project to carry out accommodation works. It is likely the project manager will be involved to manage the project team established to carry out these works. Often this team will be separate from the main project team and will comprise personnel with greater experience of operating in a finished project environment.

- Typical elements of client accommodation works for an office building would be:
  - fitting out of special areas:
    - restaurant/dining areas
    - reception areas
    - training areas
    - executive areas
    - post rooms
    - vending areas
  - installation of IT systems:
    - servers
    - User interface units (tablets/laptops, etc.)
    - telecommunications equipment
    - audiovisual and video conferencing
  - demountable office partitions:
    - furniture
    - specialist equipment
    - security systems
    - artwork and planting

Operational commissioning

The principles of client commissioning and occupation should be determined at the feasibility and strategy stage. Client commissioning (as with occupation, which usually follows on as a continuous process) is an activity predominantly carried out by the client's personnel, assisted by the consultants as required.

The objective of client commissioning is to ensure that the facility is equipped and operating as planned and to the initial concept of the business plan established for the brief. This entails the formation, under the supervision of the client's occupation coordinator, of an operating team early in the project so that requirements can be built into the contract specifications. Ideally, the operating team is formed in time to participate in the design process. (Their role is identified in the Project Handbook Briefing Note 1.04, supported by a checklist in Briefing Note 7.01.)

Main tasks

The main tasks are as follows:

- Establishing the operating and occupation objectives in time, cost, quality and performance terms. Consideration must be given to the overall implications of phased commissioning and priorities defined for sectional completions, particular areas/services and security.

- Arranging the appointment of the operating team in liaison with the client. This is done before or during the detailed design stage, so that appropriate commissioning activities can be readily included in the contract.

- Making sure at budget stage that an appropriate allowance for the client's commissioning costs is made. Accommodation schedule of works can potentially consume a significant part of the total project budget.

- Preparing role and job descriptions (responsibilities, time-scales, outputs) for each member of the operating team. These should be compatible with the construction programme and any other work demands on members of the operating team.

- Coordinating the preparation of the client's commissioning schedule and an action list in liaison with the client, using a commissioning checklist (see Briefing Note 7.01).

- Arranging appropriate access, as necessary, for the operating team and other client personnel during construction, by suitable modification of the contract documents.

- Arranging coordination and liaison with the contractors and the consultants to plan and supervise the engineering services commissioning, for example, preparation of new work practices manuals, staff training and recruitment of additional staff if necessary; the format of all commissioning records; renting equipment to meet short-term demands; overtime requirements to meet the procurement plan; meeting the quality and performance standards, all as defined in stage 6.

- Considering early appointment/secondment of a member of the client management team to act as the occupation coordinator; this ensures a smooth transition from a construction site to an effectively operated and properly maintained facility (see Briefing Note 7.02 for an introduction to facilities management).

- Before the new development can be occupied, the client needs to operationally commission various elements of the development. This involves setting to work various systems and preparing staff ready to run the development and its installations:

- transfer of technology

- checking voice and data installation are operational

- stocking and equipping areas such as a restaurant

- training staff for running various systems

- training staff to run the property

- Also part of the client's operational commissioning is the obtaining of the necessary statutory approvals needed to occupy the building, such as the occupation certificate and the environmental health officer's approval of kitchen areas (if applicable).

- Occupation of developed property depends on detailed planning of the many spaces to be used. For office buildings, this space planning process is developed progressively throughout the project lifecycle.

- Final determination of seating layouts is delayed until the occupation stage in order to accommodate the latent changes to the client's business structure. A typical space planning process consists of:

  - confirming the client's space standards including policy on open plan and cellular offices

  - confirming the client's furniture standards

  - determining departmental headcount and specific requirements

  - determining an organisational model of the client's business, reflecting the operational dependencies and affinities

  - develop a building stacking in order to fit the gross space of each department within the overall space of the building

  - develop departmental layouts to show how each department fits the space allocated to it

  - develop furniture seating layouts in order to allocate individual names to desks

- It is essential that for each of these stages the client organisation in the form of user liaison groups has a direct involvement and approves each stage.

- Moving or combining businesses into new premises is a major operation for a client. During the duration of the move, there is the potential for significant disruption to the client's business. The longer the move period, the greater the risks to the client. Migration therefore requires a significant level of planning. Often the client will appoint a manager separate from the new building project to take overall responsibility for the migration. For major or critical migrations, the client should consider the use of specialist migration consultants to support their own resources.

- During the migration planning, a number of key strategic issues need to be addressed:

  - determining how the building will be occupied

  - establishing the timing of the move

  - identifying the key activities involved in the migration and assigning responsible managers

  - determining move groups and sequence of moves to minimise business disruption

  - determining the project structure for managing the move

7 Completion, handover and operation

- identifying potential risks that could impact on the move

- involving and keeping the client's staff informed

- As some of these strategic issues could have an impact on the timing and sequencing of the main building works, it is important to address them early in the project lifecycle.

- The final part of occupation is the actual move management. This involves the appointment of a removal contractor, planning the detailed tactics of the move, and supervision of the move itself.

- The overall period that the move takes is determined by the number of items to be transferred with each member of the staff and by the degree of difficulty of transferring IT systems for each move group.

- A critical decision for the client during the occupation stage is the point at which a freeze is imposed on space planning and no further modifications are accommodated until after migration has been achieved.

- It is likely that the factor having most impact on the timing of the freeze date will be the setting up of individual voice and data system profiles.

- It would be common for clients to impose an embargo on changes both sides of the migration and for the client then to carry out a post-migration sub-project to introduce all the required changes required by departments.

Client occupation

Occupation should follow a very carefully planned logistical schedule managed by the incoming user of the facility following completion of construction. This can be put under the overarching control of the project manager or can be headed by an appointed occupation coordinator.

Unlike many other project management activities, occupation involves employees themselves and is affected by the style of management and the culture of the user's organisation. Consequently, well-executed planning and the end user's involvement in the process can result in better management/employee relations, bringing a greater feeling of participation and commitment to the workforce.

It is normal to produce an operational policy document in the planning stages which is a blueprint for implementation of the business plan. In particular, it sets out services which will be kept in house and those services which will be contracted and how they will be procured.

The arrangements for occupation and migration from one facility to another will, on many projects, be predetermined by space-planning exercises carried out in the initial design stages by the design team or space-planning consultant. The guidance given in this chapter should be put in the context of the overall planning of a client's needs for a particular facility. This will follow from:

- strategic analytical briefing

- detailed briefing (departmental level) and lead to criteria such as

- quantifying spatial requirements

- physical characteristics for each department/sector

- critical affinity groupings

- extent of amenities

- workspace standards

- office automation strategies

- security/public access

- furniture, fittings and equipment (FF&E) schedules

On complex projects, this can be taken one stage further to the production of room data sheets which form the basis of the design brief, equipment transfer or purchase, movement of personnel and facilities management.

The procedure outlined next gives a typical approach, which may need to be interpreted in order to harmonise with the practices and expectations of the users. Nevertheless, change in established practices is encouraged where doing so will smooth the process and make it more effective. Occupation can be divided into four stages as explained later and as shown in Figure 7.1, Figure 7.2, Figure 7.3 and Figure 7.4. The following services are typically outsourced on renewable annual or 3-year contracts (often provided by the landlord in managed buildings):

- reception and telephony

- security

- cleaning

- building management and operation of services and equipment

- maintenance

- IT support

- catering and waste management

- landscaping and grounds maintenance

- transport and courier services

Structure for implementation

Structure for implementation means appointment of individuals and groups to set out the necessary directions, consultation and budget/cost parameters. Figure 7.1 gives an example.

Scope and objectives

Scope and objectives means deciding what is to be done, considering the possible constraints and reviewing as necessary. Figure 7.2 gives an example.

Methodology

Methodology is how the whole process will be achieved. Identification of individual or groups of special activities and their task lists aimed at defining the parameters and other related matters, for example, financial implications. Figure 7.3 gives an example.

Organisation and control

Organisation and control means carrying out the process and keeping schedule and budget/cost under review. Figure 7.4 gives an example. The individuals and groups likely to be concerned are as follows:

- *Project executive*: appointed by the client/tenant at the director/senior management level and responsible for the complete process.

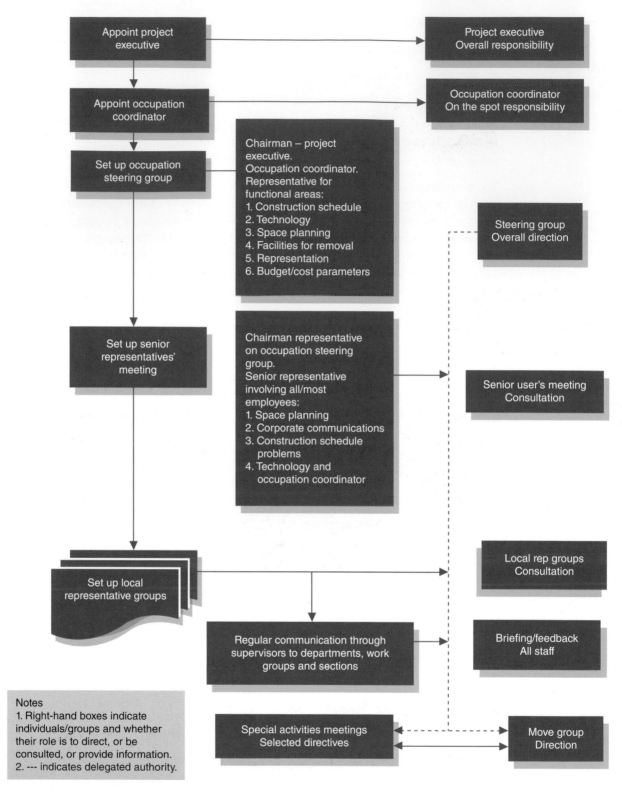

Figure 7.1 Occupation: structure for implementation.

- *Occupation coordinator*: project manager appointed, or existing one confirmed by the client, with on-the-spot responsibility.

- *Occupation steering group*: chaired by the project executive and consisting of occupation coordinator and a few selected senior representatives covering the main functional areas. Concerned with all major decisions but subject to any constraints laid down by the client, for example, financial limits.

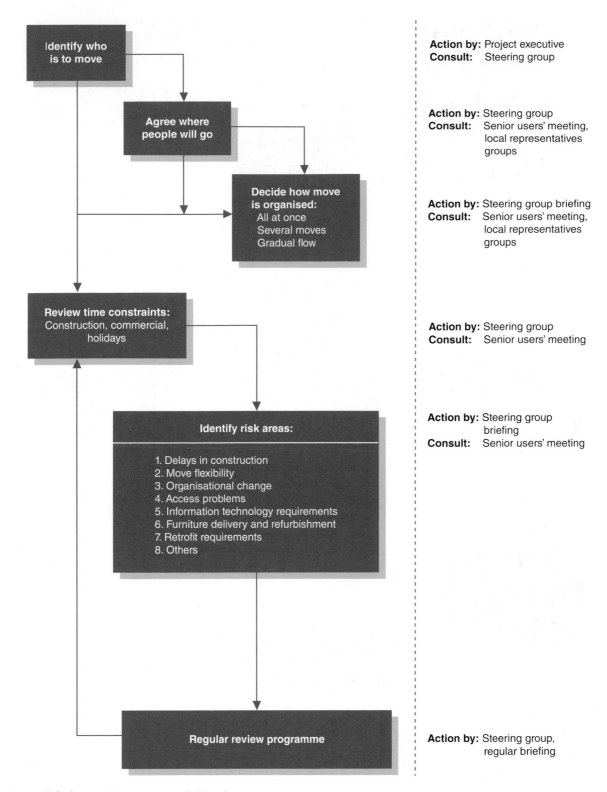

Figure 7.2 Occupation: scope and objectives.

- *Senior representatives meeting*: chaired by one of the functional representatives on the occupation steering group and made up of a few senior representatives covering the majority of employees and the occupation coordinator.

- *Local representative groups*: chaired by manager/supervisor of own group and concerned with providing views related to a particular location or department. Membership to reflect the specific interest of the group at the location.

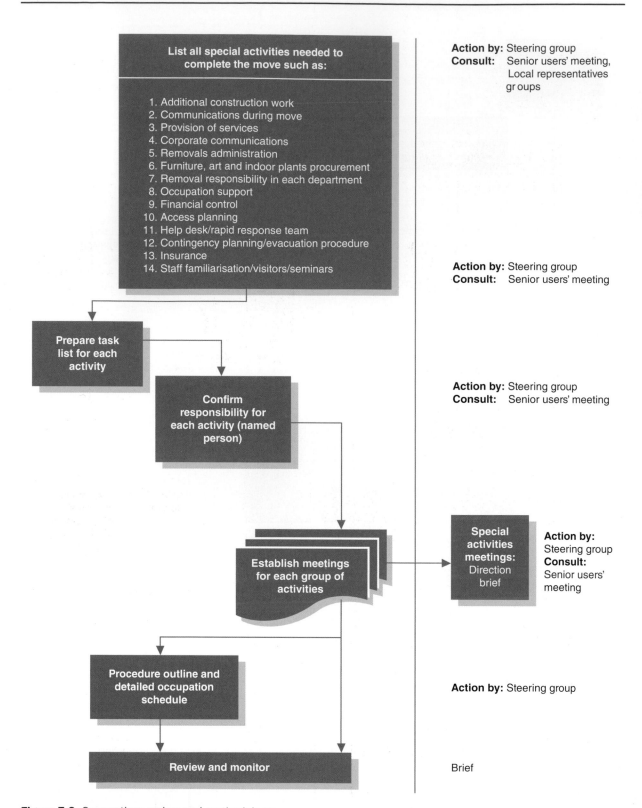

Figure 7.3 Occupation: review and methodology.

- *Special activities meetings*: meetings for individual or group of special activities as identified in 'methodology'. A single person will be made responsible for achieving all the tasks which make up a special activity and will chair the respective meetings.

- *Move group*: responsible for the overall direction of the physical move, having been delegated by the occupation steering group, the task of detailed preparation and control of the move programme including its budget/cost.

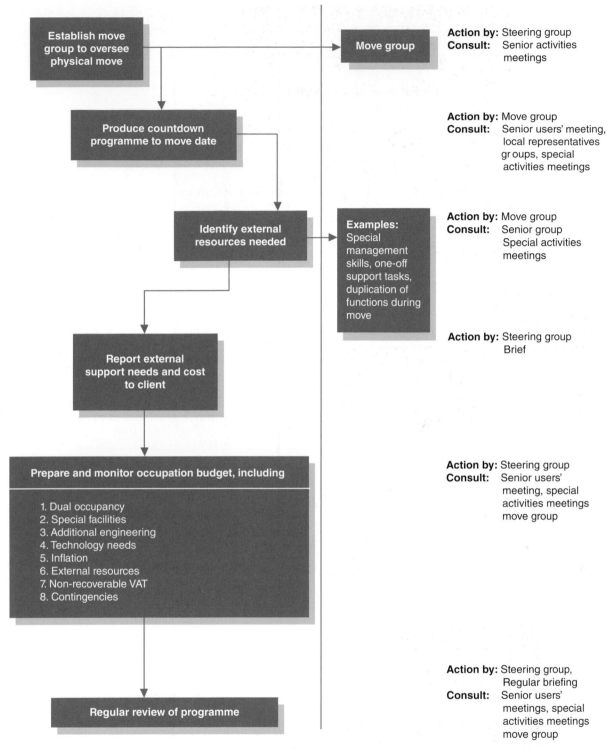

Figure 7.4 Occupation: organisation and control.

- *Briefing groups*: concerned with effective and regular communication with all employees to provide information to work groups/sections by their own managers or supervisors, so that questions for clarification are encouraged. Special briefings may also be vital, especially during the build-up to occupation.

On many projects, the individuals and groups identified earlier may be synonymous with those given under client commissioning, for example, for commissioning team read occupation steering group and vice versa.

Figure 7.1, Figure 7.2, Figure 7.3 and Figure 7.4 provide an at-a-glance summary of the occupation process and Briefing Notes 7.01 and 7.03–7.05 provide checklists for a typical control system.

Soft landings

BSRIA states that soft landings means[1] designers and constructors staying involved with buildings beyond practical completion, to assist the client during the first months of operation and beyond, to help fine-tune and de-bug the systems, and ensure the occupiers understand how to control and best use their buildings.

Some of the advantages of soft landings are stated as:

- ensuring sustainability targets are realised

- understanding what works in practice and what needs to be improved

- unified vehicle for achieving tighter environmental performance and the best opportunity for producing zero-carbon buildings that actually meet their design targets

- shifts the emphasis for good performance away from just design aspiration to the way buildings are actually managed and maintained.

- only requires small extra funding, well within the margin of competitive bids.

- the three-year aftercare period, typically involving the architect and building services engineer, does involve extra costs, but these are modest in relation to the value that's added to the client's building

BIM strategy

At completion of construction, the project team can hand over the as-built status project BIM, and data sets to the client and their Facilities Management (FM) team.

At the briefing and definition stage, it will have been determined what information the client requires at handover. Some clients will not have the capability to accept and make use of a project BIM. Therefore, they will need the handover information in traditional 2D forms. So as-built drawings, operations and maintenance manuals, CDM health and safety files and any other user/FM-orientated information.

In a BIM environment, all of these can be contained within the BIM model and datasets. If the client has the capability and systems in place, this can then be imported into the client's FM management software. FM managers can access the BIM using tablets as they walk round the building asset, and obtain any information as they need it, and also update the FM BIM, with additional information or comments.

---

[1] 'Introducing Soft Landings' – a BSRIA/Usable Buildings Trust Publication, https://www.bsria.co.uk/services/design/soft-landings/guidance/ (accessed March 2014).

Briefing Note 7.01 Client commissioning checklist

Brief	■ ensure roles and responsibilities for commissioning team are developed and understood progressively from the feasibility and strategy stages
Budget schedule	■ based upon a clear understanding and agreement of the client's objectives
Commissioning action checklist	■ investigate and identify commissioning requirements management control document
Appointments	■ commissioning team ■ operating and maintenance personnel ■ aftercare engineer ■ job descriptions, time-scales and outputs must be documented and agreed
Client operating procedures	■ work practice standards ■ health and safety at work requirements
Training of staff	■ services ■ security ■ maintenance ■ procedures ■ equipment
Client equipment (including equipment rented for commissioning)	■ schedule ■ selection ■ approval ■ delivery ■ installation

Building services and equipment	■ define/check standards required in tender specification
	  ☐ testing
	  ☐ balancing ⎫
	  ☐ adjusting ⎬ detail format of records
	  ☐ fine tuning ⎭
	■ marking and labelling, including preparation of record drawings
	  ☐ handover of spares ⎫ must be compatible with any planned
	  ☐ handover of tools ⎬ maintenance or equipment
	  ⎭ standardisation policies

Maintenance	■ acceptance by client's maintenance section from the client's construction and commissioning team ■ arrangements ■ procedures ■ contracts

(Continued)

Security	■ alarm systems
	■ telephone link
	■ staff routes
	■ access (including card access)
	■ fire routes
	■ bank cash dispensers
Communications	■ telephones
	■ radios
	■ paging
	■ public address systems
	■ easy-to-read plan of building
	■ data links
Signs and graphics	■ code of practice for the industry
	■ statutory notices - H&S, fire, Factories Act, unions
Initiation of operations	■ final cleaning
	■ maintenance procedures (including manufacturers' specialist maintenance)
	■ cleaning and refuse collection
	■ insurance required by date and extent of cover will vary with the form of contract
	■ access and security (including staff identity cards)
	■ safety
	■ meter readings or commencement of accounts for gas, water, electricity,
	■ telephone and fuel oil
	■ equipping
	■ staff 'decanting'
	■ publicity
	■ opening arrangements
Review operation of facility	■ at 6, 9 and 12 months (including energy costs)
	■ improvements and system fine-tuning
	■ defects reporting, correction and verification procedures
	■ latent defects
Feedback	■ channelled through aftercare engineer if appointed

Briefing Note 7.02　Introduction to facilities management

Facilities management (FM) started out as property management primarily concerned with the management of premises. As commercial reality and competitiveness demanded greater efficiency, attention focused on the need to manage not just the buildings but the entire resources used by organisations in the generation of their wealth, hence FM. It is not a new concept but one that has progressed from use by a handful of companies to become the fastest-growing property and resource management sector in construction.

FM seeks to create a framework that embraces the traditional estate management functions of property maintenance, lighting and heating with increasingly analytical reviews of space occupation/planning, asset registers, health and safety registers, and activity flow throughout the premises. Hence, the term facilities is used to include all the buildings, furnishings, equipment and environment available to the workforce while pursuing the company's business goals.

The success of FM has been greatly enhanced by the development of reliable and powerful computer technology together with the boom in personal computers that has made serious data handling affordable to all. The use of databases to control the occupational activities of buildings is both reactive and proactive with the latter gaining in importance. The reactive use allows data on the performance of the workplace to be collected and stored, which in turn is available for historical analysis that can be used proactively to identify recurring trends and anticipate operational problems, so eliminating waste.

Every FM application in industry and commerce is in effect a one-off system; it must address the priorities of the company but is actually assembled from a series of independent modules that operate from a universal FM platform. The emerging industry-standard platform is based on the computer-aided design technology used extensively in the design of buildings; this has been developed into powerful computer-aided facilities management (CAFM) systems. CAFM systems are increasingly likely to serve as an indispensable source of reference for the project manager and project team in drawing up the project brief for buildings of similar function. The pairing of FM and project management in this way should enable the procurement of increasingly efficient property.

The CIOB has become aware of the vast array of bespoke, tailor-made FM contracts which are prevalent in the FM industry. Many of these contracts are often based on models from other industries and suffer from lack of focus on FM contractual issues.

CIOB, in partnership with Cameron McKenna, published the first standard form of FM contract in 1999. The third edition of this document had been published in 2008.

Engineering services to be covered	■ Routinely: □ water supply and sanitation □ heating/cooling systems (boilers, calorifiers, chillers) □ ventilation systems □ air-conditioning □ electrical (generators, switchboards, others) □ mechanical (pumps, motors, others) □ fire detection and protection systems □ control systems (electrical, pneumatic, others) □ telephone/communications ■ Specialist □ process plant for food, pharmaceutical, petrochemical or manufacturing □ activities □ security (CCTV, sensors, access control) □ facility management system □ acoustic and vibration scans □ lifts, escalators, others □ IT systems and BMS
Contract documents	■ Responsibilities – client/contractor/manufacturer: □ bills of quantities/activity schedule items for commissioning activity with separate sums of clearly worded inclusion in M&E item descriptions ■ specification of commissioning: □ provision for providing that commissioning is performed - observation test results □ methods and procedures to be used, appropriate standards/codes of practice, e.g., CIBSE/IHVE/BSRIA/IEE/LPC/BS ■ provision for appropriate client access ■ client staff training ■ operating and maintenance manuals (as installed) ■ statutory approvals ■ record drawings and equipment software (as installed) and test certification ■ statutory approvals (lifts, fire protection, others) ■ insurance approvals.

Contractor's commissioning programme	■ Manufacturers' works testing
	■ site tests prior to commissioning (component testing, e.g., a fan motor)
	■ pre-commissioning checks (full system, e.g., air-conditioning, by contractor before demonstration to client)
	■ set to work (system by system)
	■ commissioning checks (including balancing/regulation)
	■ demonstration to client (system basis)
	■ performance testing (including integration of systems)
	■ post-commissioning checks (including environmental fine-tuning during facility occupancy

Briefing Note 7.04　Engineering services commissioning documents

CIBSE		
Commissioning codes	A	Air distribution
	B	Boiler plant
	C	Automatic control
	R	Refrigerating systems
	W	Water distribution systems
	TM12	Emergency lighting
BSRIA		
	TM 1/88	Commissioning HVAC systems divisions of responsibilities
	TN 1/90	European commissioning procedures
	AG 1/91	The commissioning of VAV systems in buildings
	AG 2/89	The commissioning of water systems in buildings
	AG 3/89	The commissioning of air systems in buildings
	AG 8/91	The commissioning and cleaning of water systems
	AH 2/92	Pre-commissioning of BEMS – a code of practice
	AH 3/93	Installation commissioning and maintenance of fire and security systems
HMSO		
	HTM 17	Health building engineering installations commissioning and associated activities (hospitals)
	HTM 82	Fire safety in healthcare premises fire alarms and detection systems (hospitals)
Loss Prevention Council		
	LPC	Rules for automatic sprinkler installations
IEE		
	Wiring regulations	
British Standards		
	An extensive list of BS publications exists for specialised systems and equipment, e.g. gas flues, steam and water boilers, oil and gas burning equipment, electrical equipment, earthing machinery, etc.	
IT		Cabling installation and planning guide of relevant 'equipment' manufacturer/ supplier

Handover procedure	■ design's certificate
	■ Certificate of Practical Completion
	■ health and safety file
	■ inspections and tests
	■ copies of certificates, approvals and licences
	■ release of retention monies
	■ final clean
	■ handover of spares
	■ meters read and fuel stocks
	■ final account, final inspection and final certificate arrangements
	■ liason with tenants, purchaser or financier
	■ publicity
	■ opening arrangements
	■ client's acceptance of building
	■ post-completion review/project close-out report
Schedule	■ remedial works
	■ defects liability period and defect correction
Building owner's manual	■ adjustment of building services
	■ client's fitting out
	■ consultant's contributions
	■ format
Operating and maintenance manuals, as-built drawings and C&T records	■ servicing contracts established
	■ handover to facilities manager
Letting or disposal	■ schedule
	■ publicity
	■ strategy
	■ liaison
	■ documentation
	■ insurance

(Continued)

7.05 Briefing Note

Additional works	■ contracts
	■ major service installations or adaptations
	■ fitting out
	■ shop fitting
Access by contractors	■ remedial works
	■ additional contracts
Security	■ key cabinet
	■ key schedule

Inspection Certificates and Statutory Approvals

Fire officer inspections	■ access provisions for fire tenders
	■ fire shutters
	■ fireman's lift
	■ B smoke extract system/pressurisation
	■ foam inlet/dry riser
	■ fire dampers
	■ alarm systems
	■ alarm panels
	■ telephone link
	■ fire protection systems:
	☐ sprinklers
	☐ hose reels
	☐ hand appliances/blankets, etc.
	■ statutory signs

Fire Certificate

Institution of Electrical Engineers' certificate

Water authority certificate of hardness of water

Insurer's inspections	■ fire protection systems:
	☐ sprinkler
	☐ hose reels
	☐ hand appliances
	■ lifts/escalators
	■ mechanical services:
	☐ boilers
	☐ pressure vessels
	☐ electrical services
	☐ security installations

Officers of the court inspection (licensed premises)

Pest control specialists' inspection

Environmental health officer inspection

Building control officer inspection

Planning	■ outline
	■ detailed including satisfaction of conditions
	■ listed building
Landlord's inspection	
Health and safety officer's inspection	
Crime prevention officer's inspection	
Secure by design inspection	

Project no:

Authorised to approve　　　　　　　　　　　　Signature

_____　　_____

Have/has the following been completed?

1.　Contract works. ☐
2.　Commissioning of engineering services. ☐
3.　Outstanding works schedule issued. ☐
4.　Outstanding works completed. ☐
5.　Operating and maintenance manuals, 'as-built' drawings and C&T records issued ☐
6.　Maintenance contracts put into place. ☐
7.　Building Regulations consent signed off. ☐
8.　Occupation certificate issued. ☐
9.　Public health consent signed off. ☐
10.　Health and safety consent signed off and health and safety file available. ☐
11.　Planning consent complied with in full, including reserved matters. ☐
12.　Equipment test certificates issued (lifts, cleaning cradle, others). ☐
13.　Insurers' certificates issued (lifts, cleaning cradle, sprinklers, others). ☐
14.　Means of escape signed off. ☐
15.　Fire-fighting systems and appliances signed off. ☐
16.　Fire alarm system signed off and fire certificate issued. ☐
17.　Public utilities way-leaves and lease agreements signed off. ☐
18.　Public utilities supplies inspected and signed off. ☐
19.　Licences to store controlled chemicals. ☐
20.　Licences to dispose of controlled chemicals. ☐
21.　Licences to store gases. ☐
22.　Licence to use artesian well. ☐
23.　Adoption of highways, estate roads, and walkways by local authorities. ☐
24.　Consent to erect and maintain flag-poles. ☐
25.　Consent to erect illuminated signs. ☐
26.　Cleaning to required standard. ☐
27.　Removal of unwanted materials and debris. ☐
28.　Tools and spares. ☐
29.　Client/user insurances established. ☐

Completed	/
Not applicable	x

8 Post-completion review and in use

Stage checklist

Key processes:	Post-occupancy evaluation
	Project audit
	Project feedback
	Close-out report
	Benefits realisation
	Occupation/in-use strategy
Key stage objectives:	'Has the project satisfied the *need*?'
Key deliverables:	Project close-out report
	Post-occupancy evaluation report
	Occupier's handbook
Key resources:	Client team
	Project manager
	Occupation and maintenance team

Stage process and outcomes

This, the final stage in the development process, consists of the administrative and financial closure of the project.

Outcomes:

- any post-occupancy evaluations required

- issue of the Final Certificate confirms completion of any outstanding works

- finalisation of all payments to consultants and contractors

- settlement of any claims for additional monies

- archiving of the project documentation

- post-project review

- benefits realisation appraisal (if required)

- record any residual risks/issues in Health and Safety File

- embellish and enact operation and maintenance plans

Code of Practice for Project Management for Construction and Development, Fifth Edition. Chartered Institute of Building.
© Chartered Institute of Building 2014. Published 2014 by John Wiley & Sons, Ltd.

Post-occupancy evaluation

Post-occupancy evaluation (POE[1]) is a way of providing feedback throughout a building's lifecycle from initial concept through to occupation. The information from this feedback can be used for informing future projects, whether it is on the process of delivery or technical performance of the building. The key benefits of this feedback include:

* identification of and finding solutions to problems in buildings

* response to user needs

* improve space utilisation based on feedback from use

* built-in capacity for building adaptation to organisational change and growth

* finding new uses and optimising existing uses of space

* accountability for building performance by designers

* improvements in design quality

* strategic performance review

The greatest benefits from POEs come when the information is made available to as wide an audience as possible, beyond the institution whose building is evaluated, to the whole organisation and perhaps the construction industry. Information from POEs can provide not only insights into problem resolution but also provide useful benchmark data with which other projects can be compared. This shared learning resource provides the opportunity for improving the effectiveness of building procurement where each institution has access to knowledge gained from many more building projects than it would ever complete.

An indicative outline for POE can be found in Briefing Note 8.01.

Project audit

* Brief description of the objective of the project.

* Summary of any amendments to the original project requirements and their reasons.

* Brief comment on project form of contract and other contractual/agreements provisions. Were they appropriate?

* Organisation structure, its effectiveness and adequacy of expertise/skills available.

* Master schedule: project milestones and key activities highlighting planned versus actual achievements.

* Unusual developments and difficulties encountered and their solutions.

* Brief summary of any strengths, weaknesses and lessons learned, with an overview of how effectively the project was executed with respect to the designated requirements of:

  - cost

  - scheduling

---

[1] For further detailed information see 'Guide to Post Occupancy Evaluation' – a document published by the University of Westminster, http://www.aude.ac.uk/info-centre/goodpractice/AUDE_POE_guide (accessed March 2014).

- technical competency
- quality
- health and safety aspects
- sustainability targets (environmental, social and economic)

- Was the project brief fulfilled and does the facility meet the client/user needs? What needs further modification and how could further improvements be made on a value-for-money basis?
- Indication of any improvements that could be made in future projects.

Cost and time study

- Effectiveness of:
  - cost and budgetary controls
  - claims procedures
- Authorised and final cost.
- Planned against actual costs (e.g. S-curves) and analysis of original and final budget.
- Impact of claims.
- Maintenance of necessary records to enable the financial close of the project.
- Identification of time extensions and cost differentials resulting from amendments to original requirements and/or other factors.
- Brief analysis of original and final schedules, including stipulated and actual completion date; reasons for any variations.

Human resources aspects

- Communication channels and reporting relationships (bottlenecks and their causes).
- Industrial relations problems, if any.
- General assessment and comments on staff welfare, morale and motivation.

Performance study

- Planning and scheduling activities.
- Were procedures correct and controls effective?
- Staff hours summary:
  - breakdown of planned against actual
  - sufficiency of resources to carry out work in an effective manner
- Identification of activities performed in a satisfactory manner and those deemed to have been unsatisfactory.
- Performance rating (confidential) of the consultants and contractors, for future use.

Project feedback

Project feedback necessarily reflects the lessons learnt at various stages of the project, including recommendations to the client for future projects. Ideally feedback should be obtained from all of the participants in the project team at various stages. If necessary, feedback can be obtained at the key decision-making stage (e.g. at the completion of each of the stages as outlined in this *Code of Practice*). The project feedback form should include:

- brief description of the project
- outline of the project team
- form of contract and value
- feedback on contract (suitability, administration, incentives, etc.)
- technical design
- construction methodology
- comments on the technical solution chosen
- any technical lessons to be learnt
- form of consultant appointments
- comments on consultant appointments
- project schedule
- comments on project schedule
- cost plan
- comments on cost control
- change management system
- values of changes
- major source(s) of changes/variations
- overall risk management performance
- overall financial performance
- communication issues
- organisational issues
- comments on client's role/decision-making process
- comments on overall project management including any specific issues
- other comments
- close-out report

It has to be remembered that the purpose of a project feedback is not just only to express what went wrong and why, but also to observe what has been achieved well, and if (and how) that can be improved in future projects, that is, continuous improvement.

Close-out report

The project manager should summarise the findings from the various post-completion reviews in a close-out report which is issued to the client as a formal record of the project's delivery and outcome.

Benefits realisation

Although a client responsibility, on some projects the project manager may be required to assist in the assessment of whether the intended benefits for the client's organisation have been realised.

Occupation/in-use strategy

Although a client responsibility, on some projects the project manager may be required to assist in preparation and operation of the occupation strategy which should be developed in conjunction with the end users and underpinned by the project hand over documentation. Inputs from project manager may include assistance in review of controls and performance of the plan and ensuring manuals and records are updated to reflect any changes to create a live occupier's handbook. In projects where a BIM protocol had been established, the client may wish to continue the service of the BIM manager for a smooth transition of the BIM information exchange from delivery to aftercare as highlighted next.

Client's BIM strategy

Following on from the previous stage, it is possible to use the As Built BIM to manage the operation of the completed operational built asset.

The BIM is integrated into the Client's FM software, and can also be linked to the Building Management System providing real-time interaction and monitoring between the building, its systems and the BIM.

The FM BIM can be used to manage the building, flagging up maintenance and replacement requirements as well as providing immediate access and reliable information to the Client's FM managers and end users.

However, it must be emphasised that in order to achieve effective operational use of the BIM, the requirements need to be understood and set out at the project and definition stage so that the BIM can be developed and populated with data at each stage to maximum benefit.

Briefing Note 8.01 Post-occupancy evaluation process chart

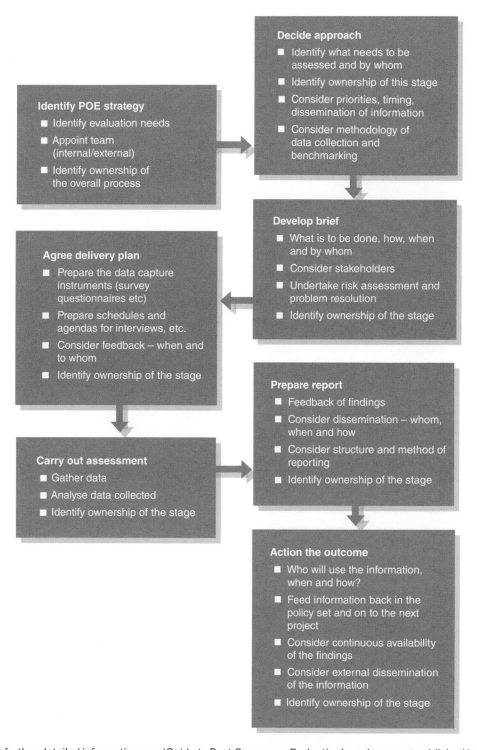

Identify POE strategy
- Identify evaluation needs
- Appoint team (internal/external)
- Identify ownership of the overall process

Decide approach
- Identify what needs to be assessed and by whom
- Identify ownership of this stage
- Consider priorities, timing, dissemination of information
- Consider methodology of data collection and benchmarking

Develop brief
- What is to be done, how, when and by whom
- Consider stakeholders
- Undertake risk assessment and problem resolution
- Identify ownership of the stage

Agree delivery plan
- Prepare the data capture instruments (survey questionnaires etc)
- Prepare schedules and agendas for interviews, etc.
- Consider feedback – when and to whom
- Identify ownership of the stage

Carry out assessment
- Gather data
- Analyse data collected
- Identify ownership of the stage

Prepare report
- Feedback of findings
- Consider dissemination – whom, when and how
- Consider structure and method of reporting
- Identify ownership of the stage

Action the outcome
- Who will use the information, when and how?
- Feed information back in the policy set and on to the next project
- Consider continuous availability of the findings
- Consider external dissemination of the information
- Identify ownership of the stage

Source: For further detailed information see 'Guide to Post Occupancy Evaluation' – a document published by the University of Westminster, http://www.aude.ac.uk/info-centre/goodpractice/AUDE_POE_guide (accessed March 2014).

Glossary

Throughout this *Code of Practice* words in the masculine also mean the feminine and vice versa. Words in the plural include the singular, e.g. 'subcontractors' could mean just one subcontractor.

Aftercare engineer	The aftercare engineer provides a support service to the client/user during the initial 6–12 months of occupancy and is, therefore, most likely a member of the commissioning team.
BIM	BIM (Building Information Modelling) enables the sharing of information and data between all stakeholders and participants around the whole asset lifecycle. It provides a platform for consistent, structured, perfect data, to enable informed smart decision making at all stages of the project process.
Budget	Quantification of resources needed to achieve a task by a set time, within which the task owners are required to work. Note: a budget consists of a financial and/or quantitative statement, prepared and approved prior to a defined period, for the purpose of attaining a given objective for that period.
Business case	Information necessary to enable approval, authorisation and policy-making bodies to assess a project proposal and reach a reasoned decision.
CDM Coordinator	A CDM Coordinator is a role required under the CDM 2007 Regulations for projects which are *notifyable* – the main duties of this role have been defied in the CDM 2007 Regulations.
Change control	A process that ensures potential changes to the deliverables of a project or the sequence of work in a project, are recorded, evaluated, authorised and managed.
Change order	An alternative name for variation order, it indicates a change to the project brief.
Client	Entity, individual or organisation commissioning and funding the project, directly or indirectly.
Client advisor	An independent construction professional engaged by the client to give advice in the early stages of a project, as advocated by the Latham Report.
Commissioning team	*Client commissioning*: predominantly the client's personnel assisted by the contractor and consultants. *Engineering services commissioning*: specialist contractors and equipment manufacturers monitored by the main contractor and consultants concerned.
Consultants	Advisors to the client and members of the project team. Also includes design team.
Contingency plan	Mitigation plan alternative course(s) of action devised to cope with project risks.

Code of Practice for Project Management for Construction and Development, Fifth Edition. Chartered Institute of Building.
© Chartered Institute of Building 2014. Published 2014 by John Wiley & Sons, Ltd.

Contractor	Generally applied to: (a) the main contractor responsible for the total construction and completion process; or (b) two or more contractors responsible under separate contractual provisions for major or high technology parts of a very complex facility. (See Subcontractor).
Design audit	Carried out by members of an *independent* design team providing confirmation or otherwise that the project design meets, in the best possible way, the client's brief and objectives.
Design delivery manager	Design delivery manager (DDM) will be typically responsible for ensuring that the design outputs are delivered to the project team and the relevant supply chain in accordance with the agreed schedule and level of detail.
Design freeze	Completion and client's final approval of the design and associated processes, i.e. no further changes are contemplated or accepted within the budget approved in the project brief.
Design management	The design management process includes the management of all project related design activities, people, processes, and resources.
Design team	Architects, engineers and technology specialists responsible for the conceptual design aspects and their development into drawings, specifications and instructions required for construction of the facility and associated processes.
Design team leader	Design teal leader (DTL) is ultimately responsible for production, coordination and adequacy of design outputs.
Development Control	Development Control (or development management in Scotland) is the element in the UK through which local government regulates land use and new construction through town & country planning.
End user	Organisation or individual who occupies and operates the facility and may or may not be the client.
Environmental management	Environmental management will include establishing and identifying environmental impact assessments and formulating an environmental statement to be adhered to by all parties throughout the lifecycle of the project.
Facilities management	Planning, organisation and managing physical assets and their related support services in a cost-effective way to give the optimum return on investment in both financial and quality terms.
Facility	All types of constructions, e.g. buildings, shopping malls, terminals, hospitals, hotels, sporting/leisure centres, industrial/processing/chemical plants and installations and other infrastructure projects.
Feasibility stage	Initial project development and planning carried out by assessing the client's objectives and providing advice and expertise in order to help the client define more precisely what is needed and how it can be achieved.
Handbook	See **Project handbook**.
KPI	KPI or Key Performance Indicator is devised to generate information on the range of performance being achieved from a measurement and monitoring perspective.
Life-cycle costing	Establishes the present value of the total cost of an asset over its operating life, using discounted cash flow techniques, for the purpose of comparison with alternatives available. This enables investment options to be more effectively evaluated for decision-making.

Master Programme	This is the name given under some forms of contract to the baseline schedule, against which progress is expected to be monitored. It bears no relationship to the concept of the dynamic working schedule, used as a time model for the purposes of time management.
Occupation	Sometimes called *migration* or *decanting*. It is the actual process of physical movement (transfer) and placement of personnel (employees) into their new working environment of the facility.
Planning	The determination and communication of an intended course of action incorporating detailed method(s) showing time, place and resources required.
Planning gain	A condition attached to a planning approval which brings benefits to the community at a developer's expense.
Post Occupancy Evaluation	Post Occupancy Evaluation (POE) is a way of providing feedback throughout a building's life cycle from initial concept through to occupation.
Principal contractor	The contractor appointed by a client under the CDM Regulations to carry out this role.
Programme management	A programme of works comprises a number of projects that are related because they contribute to a common outcome. Programme management provides co-ordinated governance to the realisation of benefits that result from projects; it is concerned with initiating projects, managing the interdependencies between projects, managing risk, and resolving conflicting priorities and resources across the projects.
Project	Unique process, consisting of a set of co-ordinated and controlled activities with start and finish dates, undertaken to achieve an objective conforming to specific requirements, including constraints of time, cost and resources.
Project brief	Statement that describes the purpose, cost, time and performance requirements/constraints for a project.
Project execution plan	A plan for carrying out a project, to meet specific objectives, that is prepared by or for the project manager. In some instances this is also known as the project management plan.
Project governance	Project governance helps make sure that a project is executed according to the standards of the organization performing the project. Governance keeps all project activities above board and ethical, and also creates accountability.
Project handbook	Guide to the project team members in the performance of their duties, identifying their responsibilities and detailing the various activities and procedures (often called the project bible). Also called project execution plan, project manual and project quality plan.
Project insurance	Project insurance is the descriptive title for a suite of insurances that are specifically designed to meet the needs of individual projects as opposed to relying on the individual insurance arrangements of the project team.
Project manager	Individual or body with authority, accountability and responsibility for managing a project to achieve specific objectives.
Project schedule	Time plan for a project or process. Note: on a construction project this is usually referred to as a 'project programme'. The construction industry tends to refer to programmes rather than schedules. Indeed the term schedule tends to mean a schedule of items in tabular form, e.g. door schedule, ironmongery schedule, etc.

Glossary

Project sponsor	The project sponsor represents the client (which is usually the government) acting as a single focal point of contact with the project manager for the day-to-day management of the interests of the client organisation.
Project team	Client, project manager, design team, consultants, contractors and subcontractors.
Risk	Combination of the probability or frequency of occurrence of a defined threat or opportunity and the magnitude of the consequences of the occurrence.
Risk analysis	Systematic use of available information to determine how often specified events may occur and the magnitude of their likely consequences.
Risk factor	Associated with the anticipation and reduction of the effects of risk and problems by a proactive approach to project development and planning.
Risk management	Systematic application of policies, procedures, methods and practices to the tasks of identifying, analysing, evaluating, treating and monitoring risk.
Risk register	Formal record of identified risks.
Seasonal commissioning	Seasonal commissioning involves re-commissioning systems affected by seasonal changes in winter and summer.
Stakeholder	Any individual or entity that has an influence on or is being impacted upon (directly or indirectly) by the project.
Strategy stage	During this stage a sound basis is created for the client on which decisions can be made allowing the project to proceed to completion. It provides a framework for the effective execution of the project.
Subcontractor	An individual or company to whom the contractor sublets the whole or any part of the works. This covers such elements as design, specialist trades and labour-only supply.
Sustainability	Sustainability in construction and development aims at reducing the environmental impact of the construction and development projects, over its entire life cycle, while optimising resource efficiency, economic viability and functional performance.
Tenant	Facility user who is generally not the client or the developer.
User	The ultimate occupier of the facility.

Bibliography

The following is not intended to provide a comprehensive guide to the vast amount of literature available. Rather it is intended to support readers by directing them to supplementary titles which will allow construction project management and the intertwined processes to be evaluated and understood within its appropriate context.

Reflecting the nature of the evolution of the subject area and the Code of Practice, the core documentation has been separated to 5th edition and 4th edition[1] publications, for ease of reference.

5th Edition Publications

Anonymous (2007) The economics of climate change. *The Stern Review*.

Eastman, C., Teicholz, P., Sacks, R. & Liston, K. (2011) *BIM Handbook A Guide to Building Information Modelling*, 2nd edn. John Wiley & Sons, Hoboken

Fewings, P. (2012) *Construction Project Management: An Integrated Approach*, 2nd edn. Routledge, London.

Goleman, D. (2000) Leadership that gets results. *Harvard Business Review*, March–April.

Johnson, G; Scholes, K and Whittington R (2006) *Exploring Corporate Strategy*, FT Prentice Hall London.

Lock, D. (2013) *Project Management*, 10th edn. Gower, Surrey.

Mead, J. & Gruneberg, S. (2013) *Programme Procurement in Construction: Learning from London 2012*, Wiley-Blackwell, Oxford.

Mike Jacka, J. & Keller, P.J. (2009) *Business Process Mapping: Improving Customer Satisfaction*. p. 257. John Wiley & Sons, New York.

Morgan, A. & Gbedemah, S. (2010) How poor project governance causes delays. A paper presented to the *Society of Construction Law at Meeting in London*, 2 February 2010.

Morris, P. (2013) *Reconstructing Project Management*. Wiley-Blackwell, Oxford.

Office of Government Commerce (2005) Common causes of project failure, http://www.dfpni.gov.uk/content_-_successful_delivery-newpage-50 (accessed March 2014).

---

[1] For OGC publications refer to The National Archives.

Pickavance, K. (Spring 2005) Dispute resolution without tears, *Times of the Islands*.

Ritz, G.J. & Levy, S.M. (2013) *Total Construction Project Management*, 2nd edn. McGraw-Hill Education, New York.

Shackleton, V. (1995) *Business Leadership*. Routledge, London.

Smith, N.J. (2002) *Engineering Project Management*, 2nd edn. Blackwell Science, Ames.

The Chartered Institute of Building (2013) *Design Manager's Handbook*. Wiley Blackwell, Oxford.

Underwood, J. & Khosrowshahi, F. (2012) ITC expenditure and trends in the UK construction industry in facing the challenges of the global economic crisis. *Journal of Information Technology in Construction*, 17, 25–42, http://www.itcon.org/2012/2 (accessed April 2014).

United Nations (1987) *Report of the World Commission on Environment and Development*.

4th Edition Publications[2]

A Guide to Managing Health and Safety in Construction (1995). Health and Safety Executive Books.

A Guide to Project Team Partnering (2002). Construction Industry Council.

A Guide to Quality-based Selection of Consultants: A Key to Design Quality. Construction Industry Council.

Accelerating Change – Rethinking Construction (2002). Strategic Forum for Construction.

ACE Client Guide (2000). Association of Consulting Engineers.

Achieving Excellence through Health and Safety. Office of the Government Commerce.

Adding Value through the Project Management of CDM (2000). Royal Institute of British Architects.

APM Competence Framework (2008). Association for Project Management High Wycombe.

APM Introduction to Programme Management (2007). Association for Project Management – PMSI Group, High Wycombe.

Appointment of Consultants and Contractors. Office of the Government Commerce.

Benchmarking. Office of the Government Commerce.

Bennett, J. (1985) *Construction Project Management*. Butterworth, London.

Bennett, J. & Peace, S. (2006) *Partnering in the Construction Industry – A Code of Practice for Strategic Collaborative Working*. CIOB/Butterworth Heinemann.

Bennis, W.G. & Nanus, B. (1985) *Leaders: The Strategies for Taking Charge*. Harper & Row, New York.

Best Value in Construction (2002). Royal Institution of Chartered Surveyors

---

[2] Publications cited reflect the current publications at the time of the 4th Edition.

Briefing the Team (1996). Construction Industry Board.

Building a Better Quality of Life, A Strategy for More Sustainable Construction (2000). Department of Environment, Transport and the Regions/Health and Safety Executive.

Burke, R. (2001) *Project Management Planning and Control Techniques*, 3rd edn.

Client Guide to the Appointment of a Quantity Surveyor (1992). Royal Institution of Chartered Surveyors.

Code of Estimating Practice, 7th edn (2009). The Chartered Institute of Building.

Code of Practice for Selection of Main Contractors (1997). Construction Industry Board.

Code of Practice for Selection of Subcontractors (1997). Construction Industry Board.

Constructing Success: Code of Practice for Clients of the Construction Industry (1997). Construction Industry Board.

Sir Michael Latham (1994) *Constructing the Team*. Final report of the Government/ industry review of procurement and contractual arrangements in the UK construction industry (the Latham Report), HMSO.

Construction (Design and Management) Regulations (2007). Health and Safety Executive.

Construction Best Practice Programme (CBPP) Fact Sheets.

Construction Management Contract Agreement (Client/Construction Manager) (2002). Royal Institute of British Architects.

Construction Management Contract Guide (2002). Royal Institute of British Architects.

Construction Project Management Skills (2002). Construction Industry Council.

Control of Risk – A Guide to the Systemic Management of Risk from Construction (SP 125) (1996). Construction Industry Research and Information Association.

Cox, A. & Ireland, P. (2003) *Managing Construction Supply Chains*. Thomas Telford, London.

Dallas, M. (2006) *Value & Risk Management – A Guide to Best Practice*. CIOB/ Blackwell, Oxford.

Earned Value Management: APM Guidelines (2008) Association for Project Management – EVMSI Group, High Wycombe.

Essential Requirements for Construction Procurement Guide. Office of the Government Commerce

Essentials of Project Management (2001). Royal Institute of British Architects.

Fielder, F.E. (1967) *A Theory of Leadership Effectiveness*. McGraw-Hill, New York.

Financial Aspects of Projects. Office of the Government Commerce.

Goleman, D. (2000) Leadership that gets results. *Harvard Business Review*, March–April.

Good Design Is Good Investment. Advice to Client, Selection of Consulting Engineer, and Fee Competition (1991). Association of Consulting Engineers.

Gray, C. (1998) *Value for Money*. Thomas Telford, London.

Green, D. (ed.) (2000) *Advancing Best Value in the Built Environment – A Guide to Best Practice*. Thomas Telford, London.

Guide to Good Practice for the Management of Time in Complex Projects (2011). Chartered Institute of Building.

Guide to Project Management BS 6079 – 1 (2000). British Standards Institution.

Hamilton, A. (2001) *Managing Projects for Success*. Thomas Telford, London.

Interfacing Risk and Earned Value Management (2008). Association for Project Management – RSI Group, High Wycombe.

Kotter, J. (1990) *A Force for Change: How Leadership Differs from Management*. Free Press, New York.

Langford, D., Hancock, M.R., Fellows, R. & Gale, A.W. (1995) *Human Resources Management in Construction*. Longman, Harlow.

Lock, D (2001) *Essentials of Project Management*. Gower Publishing.

Management Development in the Construction Industry – Guidelines for the Construction Professionals, 2nd edn (2001). Thomas Telford, London.

Managing Health and Safety in Construction. Construction (Design and Management) Regulations 1994. Approved Code of Practice and Guidance (2001). HSG224 HSE Books, Health and Safety Executive.

Managing Project Change – A Best Practice Guide (C556) (2001). Construction Industry Research and Information Association.

Manual of the BPF System for Building Design and Construction (1983). British Property Federation.

Mintzberg, H (1998) Covert leadership: notes on managing professional. *Harvard Business Review* Nov–Dec, pp. 140–147.

Models to Improve the Management of Projects (2007). Association for Project Management, High Wycombe.

Modernising Construction: Report by the Comptroller and Auditor General (2001). HMSO.

Modernising Procurement: Report by the Comptroller and Auditor General (1999). HMSO.

Morris, P.W.G. (1998) *The Management of Projects*. Thomas Telford, London.

Murdoch, I. & Hughes, W. (1992) *Construction Contracts: Law and Management*. E & FN Spon, London.

Murray-Webster, R. & Simon, P. (2007) *Starting Out in Project Management*, 2nd edn. APM, High Wycombe.

Northhouse, P. (1997) *Leadership – Theory and Practice*. Sage, Thousand Oaks.

Partnering in the Public Sector – A Toolkit for the Implementation of Post-award, Project Specific Partnering on Construction Projects (1997). European Construction Institute.

Partnering in the Team (1997). Construction Industry Board.

Planning: Delivering a Fundamental Change (2000). Department of Environment, Transport and the Regions.

Potts, K. (1995) *Major Construction Works: Contractual and Financial Management*. Longman, Harlow.

Prioritising Project Risks (2008). Association for Project Management – RSI Group, High Wycombe.

Procurement Strategies. Office of the Government Commerce.

Project Evaluation and Feedback. Office of the Government Commerce.

Project Management (2000). Royal Institute of British Architects.

Project Management Body of Knowledge, 5th edn. (2006). Association for Project Management.

Project Management in Building, 2nd edn. (1988). The Chartered Institute of Building.

Project Management Memorandum of Agreement and Conditions of Engagement (xxxx). Project Management Panel, RICS Books.

Project Management Planning and Control Techniques (2001). Royal Institute of British Architects.

Project Management Skills in the Construction Industry (1996). Construction Industry Council.

Project Risk Analysis and Management Guide, 2nd edn. (2004). Association for Project Management – RSI Group, High Wycombe.

Quality Assurance in the Building Process (1989). The Chartered Institute of Building.

Rethinking Construction – Report of the Construction Task Force to the Deputy Prime Minister on the Scope for Improving the Quality and Efficiency of UK Construction (the Egan Report) (1998). Department of Environment, Transport and the Regions.

Risk Analysis and Management for Projects (1998). Institution of Civil Engineers and Institute of Actuaries.

Safety in Excavations (Construction Information Sheet No. 8). Health and Safety Executive.

Selecting Consultants for the Team (1996). Construction Industry Board.

Selecting Contractors by Value (SP 150) (1998). Construction Industry Research and Information Association.

Shackleton, V. (1995) *Business Leadership*. Routledge, London.

Sustainability and the RICS Property Lifecycle (2009). RICS Books.

Teamworking, Partnering and Incentives. Office of the Government Commerce.

The CIC Consultants' Contract Conditions, Scope of Services and Scope of Services Handbook (2007). RIBA Publishing.

The Procurement of Professional Services: Guidelines for the Application of Competitive Tendering (1993). Thomas Telford, London.

The Procurement of Professional Services: Guidelines for the Value Assessment of Competitive Tenders (1994). Construction Industry Council.

Thinking About Building? Independent Advice for Small and Occasional Clients. Confederation of Construction Clients.

Thompson, P. & Perry, J.G. (1992) *Engineering Construction Risks – A Guide to Project Risk Analysis and Risk Management*. Thomas Telford, London.

Tichy, N & Devanna, M (1986). *The Transformational Leader*.

Turner, J.R. (1999) *The Handbook of Project-based Management*. McGraw-Hill.

Value by Competition (SP 117) (1994). Construction Industry Research and Information Association.

Value for Money in Construction Procurement. Office of the Government Commerce.

Value Management in Construction: A Client's Guide (SP 129) (1996) Construction Industry Research and Information Association.

Walker, A. (2002) *Project Management in Construction*. Blackwell, Oxford.

Whole Life Costs. Office of the Government Commerce.

Past working groups of *Code of Practice for Project Management*

Fourth Working Group for the Revision of the *Code of Practice for Project Management*

Saleem Akram BSc Eng (Civil) MSc (CM) PE MASCE MAPM FIE FCIOB	Director, Construction, Innovation and Development, CIOB
Alan Crane CBE CEng FICE FCIOB FCMI	Chair, Working Group, Vice President CIOB
Roger Waterhouse MSc FRICS FCIOB FAPM	Vice Chair; Working Group, Royal Institution of Chartered Surveyors Association for Project Management
Neil Powling DipBE FRICS DipProjMan(RICS)	Royal Institution of Chartered Surveyors
Gavin Maxwell-Hart BSc CEng FICE FIHT MCIArb FCIOB	Institution of Civil Engineers
John Campbell BSc (Hons), ARCH DIP AA RIBA	Royal Institute of British Architects
Martyn Best BA Dip Arch RIBA MAPM	Royal Institute of British Architects
Richard Biggs MSc FCIOB MAPM MCMI	Construction Industry Council
Paul Nash MSc MCIOB	Trustee of CIOB
Ian Caldwell BSc BARch RIBA ARIAS MCMI MIOD	
Professor James Somerville FCIOB MRICS MAPM MCMI PhD	
Professor John Bennett DSc FRICS	
David Woolven MSc FCIOB	
Artin Hovsepian BSc (Hons) MCIOB MASI	
Eric Stokes MCIOB FHEA MRIN	
Dr Milan Radosavljevic UDIG PhD MIZS-CEng ICIOB	
Arnab Mukherjee BEng(Hons) MSc (CM) MBA MAPM MCIOB	Technical editor

The following also contributed in development of the fourth edition of the *Code of Practice for Project Management*

Keith Pickavance	Past President, CIOB
Howard Prosser CMIOSH MCIOB	Chair, Health & Safety Group, CIOB
Sarah Peace BA (Hons) MSc PhD	Research Manager, CIOB
Mark Russell BSc (Hons) MCIOB	Co-ordinator, Time Management Group, CIOB
Andrzej Minasowicz DSc PhD Eng FCIOB PSMB SIDiR	Vice Director of Construction Affairs, Institute of Construction Engineering and Management, Civil Engineering Faculty, Warsaw University of Technology
John Douglas FIDM FRSA	Englemere Ltd
Dr Paul Sayer	Publisher, Wiley-Blackwell, John Wiley & Sons Ltd, Oxford

Third Working Group for the Revision of the *Code of Practice for Project Management*

F A Hammond MSc Tech CEng MICE FCIOB MASCE FCMI	Chairman
Martyn Best BA Dip Arch RIBA	Royal Institute of British Architects
Alan Howlett CEng FIStructE MICE MIHT	Institution of Structural Engineers
Gavin Maxwell-Hart BSc CEng FICE FIHT MCIArb	Institution of Civil Engineers

Past working groups

Roger Waterhouse MSc FCIOB FRICS MSIB FAPM   Royal Institution of Chartered Surveyors and Association for Project Management

Richard Biggs MSc FCIOB MAPM MCMI   Association for Project Management

John Campbell   Royal Institute of British Architects

Mary Mitchell   Confederation of Construction Clients

Jonathan David BSc MSLL   Chartered Institution of Building Services Engineers

Neil Powling FRICS DipProjMan (RICS)   Royal Institution of Chartered Surveyors

Brian Teale CEng MICBSE DMS

David Trench CBE FAPM FCMI

Professor John Bennett FRICS DSc

Peter Taylor FRICS

Barry Jones FCIOB

Professor Graham Winch PhD MCIOB MAPM

Ian Guest BEng

Ian Caldwell BSc Barch RIBA ARIAS MIMgt

J C B Goring MSc BSc (Hons) MCIOB MAPM

Artin Hovsepian BSc (Hons) MCIOB MASI

Alan Beasley

David Turner

Colin Acus

Chris Williams DipLaw DipSury FCIOB MRICS FASI

Saleem Akram B Eng MSc (CM) PE FIE MASCE MAPM MACost E   Secretary and Technical Editor of third edition

Arnab Mukherjee B Eng MSc (CM)   Assistant Technical Editor

John Douglas   Englemere Ltd

David Woolven MSc FCIOB

First and Second Working Groups of the *Code of Practice for Project Management*

F A Hammond MCs Tech CEng MICE FCIOB MASCE FIMgt   Chairman

G S Ayres FRICS FCIArb FFB   Royal Institution of Chartered Surveyors

R J Cecil DipArch RIBA FRSA   Royal Institute of British Architects

D K Doran BSc Eng DIC FCGI CEng FICE FIStructE   Institution of Structural Engineers

R Elliott CEng MICE   Institution of Civil Engineers

D S Gillingham CEng FCIBSE   Chartered Institution of Building Services Engineers

R J Biggs MSc FCIOB MIMgt MAPM   Technical editor of second edition and Association for Project Management

J C B Goring BSc (Hons) MCIOB MAPM

D P Horne FCIOB FFB FIMgt

P K Smith FCIOB MAPM

R A Waterhouse MSc FCIOB MIMgt MSIB MAPM

S R Witkowski MSc (Eng)   Technical editor of first edition

P B Cullimore FCIOB ARICS MASI MIMgt   Secretary

For the second edition of the *Code* changes were made to he working group which included:

L J D Arnold FCIOB
P Lord AA Dipl (Hons) RIBA PPCSD FIMgt (replacing R J Cecil,   Royal Institute of British Architects
deceased)
N P Powling Dip BE FRICS Dip Proj Man (RICS)   Royal Institution of Chartered Surveyors
P L Watkins MCIOB MAPM   Association for Project Managers

Index

Code of Practice for Project Management for Construction and Development, Fifth Edition. Chartered Institute of Building.
© Chartered Institute of Building 2014. Published 2014 by John Wiley & Sons, Ltd.